INTENSIVE SHEEP MANAGEMENT

BY THE SAME AUTHOR

On Being a Tenant Farmer

DEDICATION

This book is dedicated to the memory of Professor Robert Boutflour CBE, MSc, Principal of the Royal Agricultural College, Cirencester, from 1931 until 1958. 'Bobby', as he was known with the greatest affection by all who knew him, and by many who did not, had the greatest influence on my life in farming. I had the privilege to be one of his students and, later, to work with him. I shall always be grateful to him for his inspiration and for his kindness to me personally. I hope that he would have approved of this book.

INTENSIVE SHEEP MANAGEMENT

HENRY R. FELL
FRAgS, NDA, MRAC

FARMING PRESS

First published 1979

Second Impression 1980

Second Edition 1985

Reprinted (with amendments), 1988

ISBN 0 85236 152 1

British Library Cataloguing in Publication Data

Fell, Henry R.
Intensive sheep management.—2nd ed.
1. Sheep—Great Britain
I. Title
636.3′08′0941 SF375.5.G7

ISBN 0-85236-152-1

Published by Farming Press Books
4 Friars Courtyard, 30–32 Princes Street
Ipswich IP1 1RJ, United Kingdom

Distributed in North America by Diamond Farm Enterprises, Box 537, Alexandria Bay, NY 13607, USA

Printed and bound in Great Britain by Page Bros (Norwich) Ltd.

CONTENTS

ACKNOWLEDGMENTS

THE AUTHOR'S NAME appears on the flysheet, and if credit is due, to him it is given. Any author knows full well, however, that he is but one among many who have made the work what it is. To all those who have so willingly, over the years, helped, inspired and supported me, I say a very sincere word of thanks.

In particular, I would mention those who have worked for me at Worlaby. Whether as foreman, shepherds, or tractor drivers, without their hard work and loyalty, nothing would have been achieved. In these times of industrial insanity, we, in farming, are indeed fortunate.

Then, a book has to be typed. That is quite a job, even if my handwriting were good, which it is not. The fact is that it was done cheerfully on top of her normal work load by my secretary, Mrs Jean Winstanley. To her, a special thank you.

And most of all to my wife, who has supported a most difficult husband with great patience and love.

ILLUSTRATIONS

PHOTOGRAPHS

FOREWORD

by Professor J. M. M. CUNNINGHAM, CBE, B Sc, PhD, FI Biol, FRSE,
Principal, West of Scotland Agricultural College

OVER SEVERAL decades agriculture has been one of our most successful industries. This can be attributed partly to the acceptance of new ideas, techniques and methods, all of which have contributed to improvements in efficiency and productivity. In this regard sheep production has tended to be less progressive so that it has become increasingly less competitive and has diminished in importance, particularly in lowland areas.

Sheep farming has been much in need of men who know and accept those traditional practices which should endure, identify those which should be modified and recognise those which should be discarded.

This book is the testimony of such a man. It is a lucid distillation of the knowledge and experience of a farmer who has advanced the art, science and practice of sheep husbandry. Presented here is an account of the difficulties and problems of intensifying sheep production, of the solutions found, the systems developed and success achieved. Contributing to this have been new ideas and practices in breeding, grazing management, disease control, forage use and feeding of sheep. Effective systems are dependent on deploying the skills and qualities of the right men and the appropriate equipment with efficient organisation throughout the annual cycle of events all towards satisfying a chosen market. This authoritative analysis of sheep production illustrates why intensification is no longer a theoretical abstraction but is a practical possibility yielding returns competitive with other farm enterprises.

Coupled with an account of the British sheep industry, its structure, systems of production and role, is a look beyond our shores to remind us of the international significance of sheep. Since sheep meat is far from being in surplus in the European Economic Community, and wool will inevitably be increasingly valued as artificial fibres increase in cost, the outlook for sheep farming in Great Britain is undoubtedly more optimistic.

This book should become part of the standard equipment of

those already in sheep production. It should be compulsory reading for those starting a sheep enterprise and will illuminate the thinking and present a vision of the potential of sheep which others have neither realised nor considered exists.

Henry Fell has already had a considerable impact on contemporary thought and practice in sheep husbandry and this book will deservedly extend it. His mentor of earlier days – Bobby Boutflour – would have been immeasurably delighted.

PREFACE TO SECOND EDITION

NO AUTHOR surely can fail to be flattered when his book meets a demand that takes it into a second edition. So six years on from the first conception I have now been through it with a fine-toothed comb in an attempt to bring my thinking up to date. I hope the result will continue to interest and to stimulate.

Many things have changed since 1979; most notably the political and economic climate in which we farmers work. Then we were still being encouraged to increase production both by public exhortation and by financial help in the form of generous funding of research, of grant aid and of market support. How long ago that now all seems! Today, milk quotas; tomorrow, certainly some form of severe restriction on cereal production. The talk is of the need to take some two million acres out of production. Few people are prepared to offer constructive suggestions as to what those acres should be used for.

So there are precious few optimists to be found in British farming in 1985. What about our sheep? Still substantially under-supplying our own market, we ought surely to be saying that here is the opportunity. A sensible and cost-conscious expansion to fill the gaps left by the cutbacks in cereals and milk production. The prospect is certainly there but it would be naive indeed to pretend that there were no danger signals flashing. Far too much of our return coming in the form of subsidy from Brussels; and all the signs of changing consumer tastes leading to a serious fall in consumption.

1985 certainly sees the British sheep industry delicately balanced! More so than most of those within it are aware of. Complacency resulting from high returns in recent years is linked to the stifling traditionalism which struck my son so forcibly on his return from a Nuffield Scholarship tour of Australia and New Zealand. These are the dangers, made all the worse by a subsidy system which fails to penalise what the market does not want.

Despite all this, I remain an optimist. We shall pass through the difficulties that are undoubtedly in front of us during the latter half of this decade. Out of them will, I am sure, come a more balanced agriculture than we have developed over the past twenty years. One hopefully with a lower cost structure and one where sheep will find their rightful place in the husbandry of our land.

'I thanke God and ever shall,
It is the Shepe hath payed for all'.

This inscription was engraved in 1485, upon his dining room window by one Master John Barton, Merchant of the Staple of Calais and Squire of Holme by Newark. A self-made man, his sheep had brought him wealth and prosperity. They enabled him to build his many gabled stone Manor House and to extend and furnish his local Parish Church, to great effect.

Chapter 1

INTRODUCTION

I LIKE SHEEP.

I make this rather simple statement of fact for a number of reasons, two of which I think are particularly important.

Firstly, I doubt very much if it is possible to make a success of farming sheep if you don't like them. I can imagine that both pigs and poultry, where the rules are fairly clearly spelt out, can be farmed in an industrial and impersonal sort of way. Indeed the chances of making much money out of them almost certainly depend upon just that kind of management. The farmer becomes an investor and a manager and, provided he is good at it, can succeed without having to admit that he enjoys the company of his sows or his hens.

Sheep, however, are not like that at all. They are just not suited to being industrialised and that is not their role, be it within British farming, or in France, or America, or in New Zealand. This book is basically about the intensification of sheep management and how, in my experience, it is possible to use the sheep as a profitable component of lowland farming, even under today's high cost conditions. But intensification and industrialisation are not the same thing at all.

We are after all talking about the production of meat; wool as well, of course, but always in a minor role. And if meat were to be our sole interest, then it is hard to argue a case for the ewe against the sow. For the sow has 16 teats and is capable of producing an average of 22 pigs a year which will convert food into meat at less than 3 to 1. All our efforts at breeding better sheep can hardly make the ewe compete with that performance!

THE ROLE OF THE EWE

No, the role of the ewe as a producer of meat must remain firmly where it always has been – on the land as a consumer and converter of forage. This could be forage in the form of heather and natural grasses on top of a mountain, or in the form of arable by-products such as sugar beet tops on a Lincolnshire arable farm, or forage from a first-class highly productive ley. And the management of livestock under these conditions just cannot be reduced to following a set of rules. That indefinable something which we call stockmanship – that ability to smell trouble before it comes; that instinct which leads you to do the right thing today which may have been the wrong thing last year – really can't exist unless you love the brutes!

Secondly, I like sheep because I am after all a Lincolnshire arable farmer. In the eyes of many who may read this book, I must be presumed to be more interested in the broad acres of my adopted county covered in wheat and sugar beet, peas and potatoes. Interested I certainly am, and very deeply at that, for much of my income comes from such crops. Their successful cultivation demands not only my skill and attention, but also my affection. For I sincerely believe that to succeed in farming the soil, affection for it is an essential part of one's equipment. But Lincolnshire was not always the land of crops, bereft of livestock. Time was when the great flocks of folded sheep occupied the dominant place that the pea, the potato and the sugar beet occupy today. One doesn't have to be a romantic to believe that sheep still have an important role to play in lowland farming and that is one of the themes of this book.

I first became really deeply interested in sheep in 1960. I had taken the tenancy of Worlaby in 1959 – a farm with something of a reputation for being really difficult: low lying, heavy clay and desperately wet. There never was any argument about the first priority – drainage. And drain it we did, and I know of few more satisfying things in life than to see water coming out of a drain, drying the water out of land, turning the unproductive into production.

IMPORTANCE OF GRASS

Drainage of wet land is dramatic, making the unfarmable farmable, but it is far from being the end of the story. Heavy land is never easy to farm, but a major part of any success is the maintenance of good soil structure – by well-timed cultivations, using the weather, particularly frost, but, equally important, maintaining organic matter at a reasonably high level. And this was where my interest in sheep really started – with grass. I became convinced that to farm Worlaby successfully in the long term, grassland in the rotation was essential so as to build up and maintain a good soil structure.

So I started with the conviction that Worlaby needed grass, but how to use it was a question about which I felt far less confident. Previous occupiers had produced milk, but Worlaby is not really suitable for dairying; it is nearly two miles long with all the buildings and services at one end. In any case I hadn't got that sort of capital available. Beef production? There was no prospect of making any money in beef. Grass drying? Well, that didn't look very attractive either, and in retrospect, thank goodness I wasn't tempted.

So I was left with sheep. The Cinderella of the British livestock industry, with precious little in the way of research or development going on to show some way forward. And with the market dominated by low-priced New Zealand imports. At first sight, not a very attractive prospect. But the challenge was there. Was it not possible to devise some way of modernising the basic idea of the folded arable flock? For in its heyday the folded flock was intensive arable sheep management par excellence. It had been finally abandoned in the difficult days of the 1930s, largely because it hadn't adapted to different markets and economic conditions. But as a contributor to soil fertility and as a producer of high output per acre, it would be hard to beat.

Then there was the challenge and the interest to be found in breeding. I have never wanted to be simply a producer of lambs for the slaughterhouse. There were deeper satisfactions to be found in breeding for improvement; and with some notable exceptions, such as Oscar Colburn of Northleach, it seemed to me that little constructive work was going on. Breeding was almost entirely associated with the show ring, and selection was based upon sheep which were often overfed and dressed up, rather than on measurement of economic performance. I felt that there was much to do; and I still do.

So I was attracted to sheep, but the fundamental objective

remained – to use the grass and the sheep on it as an integral and important part of the farm's arable cropping. The fact that the grass would contribute to arable profits was, however, never any reason why it shouldn't be profitable in its own right.

In the years that have gone by since 1960, we have been able to meet these objectives. In the process I have learnt a very great deal: a lot from other people; much from my own mistakes. Some of it is perhaps worth passing on.

Chapter 2

THE STRATIFICATION OF THE BRITISH SHEEP INDUSTRY

As far as I am aware, Britain is unique in the way in which its sheep industry is structured. Perhaps structured is the wrong word to use, in that it suggests something deliberately built up. This is not the case. And the fact that this inter-relationship between sheep breeding and production at different levels and in different areas has come about almost accidentally should not in any way detract from its importance. For it gives us significant economic advantage over many of our competitors; and we should be seeking ways and means of making the best use of it and maybe even improving it.

In my view, this is of such fundamental importance that it merits detailed consideration. Expressed in its simplest form, it is a relationship between three different geographic levels of sheep production, each contributing different breeds and different qualities which combine together to provide the best sheep solution for any given set of circumstances.

This is perhaps best illustrated in the form of a diagram (Fig. 1 – p.23).

There is this movement of sheep downhill making use of pure-breeding and cross-breeding to produce different 'types' of sheep for different environments. It gives us a number of significant advantages and also two particularly rewarding 'points of attack' in any genetic improvement programme. Of course, it is far from being as simple as all that; it hardly could be in an industry as complex as ours, and there is a risk of over-simplification. Not all breeds fit into tidy categories; not all draft hill ewes are crossed with Longwool rams, and when

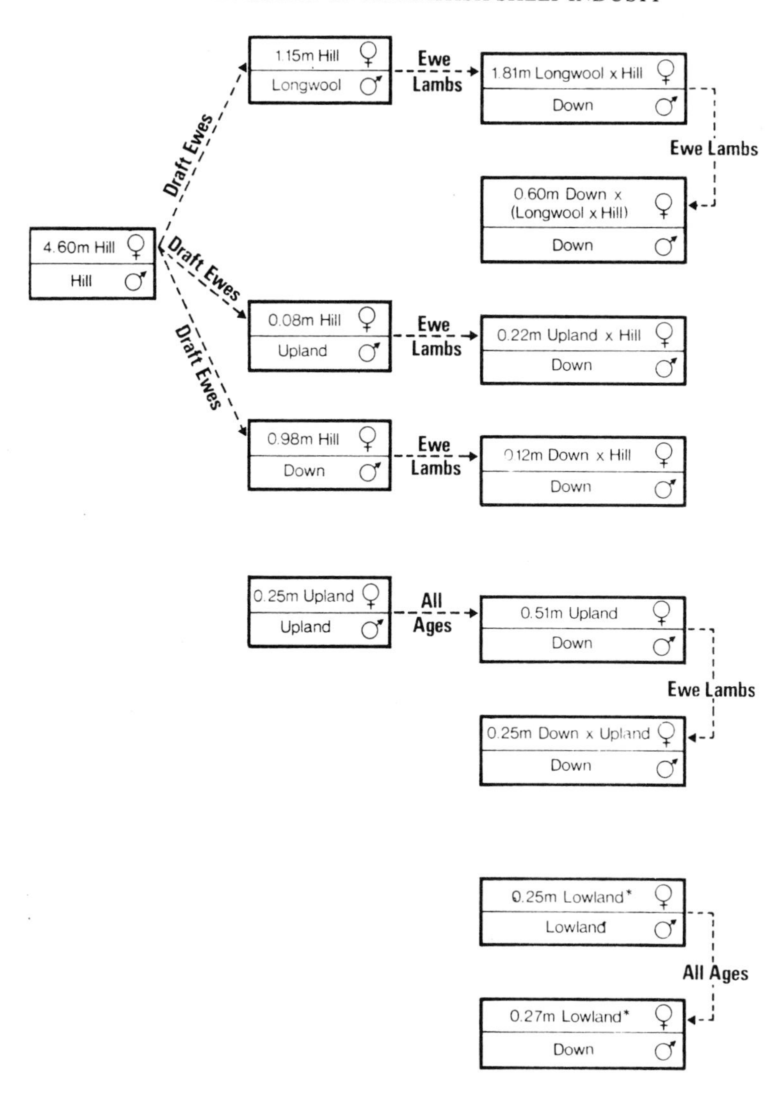

Fig. 1. Breeding structure of sheep industry in Great Britain, 1972: diagrammatic representation of relationship between main pure-bred and cross-bred groups. No survey of breed structure has been carried out since 1972. Numbers have increased since then to a total of 15·5 million breeding ewes, but the proportions between each category remain broadly the same.

they are, not all their ewe lambs end up as half-bred ewes in the lowlands; and not all lowland ewes are half-breeds. And yet the basic system, as it has evolved, is simple and efficient.

Table 1. Changes in the British Ewe Flock 1971–83

	Ewes (000)				*Shearling Ewes (000)*			
	1971		*1983*		*1971*		*1983*	
	No.	*%*	*No.*	*%*	*No.*	*%*	*No.*	*%*
England	4,401	43·9	5,964	46·8	879	40·3	1,265	44·9
Scotland	3,142	31·3	3,317	26·1	702	32·2	756	26·9
Wales	2,485	24·8	3,447	27·1	601	27·5	793	28·2
	10,028	100	12,728	100 (+26·9)	2,182	100	2,814	100 (+28·9)

FROM THE HILLS TO THE LOWLANDS

Before going on to discuss the advantages of this stratification and where we may look for improvement, we ought just to set out the background of each of the three geographic areas.

The Hills

Of the 15·5 million breeding ewes in Britain, more than half belong to the recognised hill breeds. Kept under conditions of considerable severity, they are extremely hardy, and considering the circumstances, remarkably productive. They represent a national resource of great value, and we are fortunate that national agricultural policy has kept this resource in good working order, if not at its full potential. They are the only ways of using large areas of poor land for food production, but they are far more than that – for they are the genetic base on which our prolific lowland half-bred flocks are built. Under their natural hill conditions they are, of course, bred pure, and are subjected by the environment, if not by the breeder, to a degree of selection. There can be no question of 'improving' the hill breeds by introducing more productive and therefore softer blood, simply because the resulting cross-bred would not survive.

If hill sheep farming stopped there, then that would just be an interesting form of hill land management, but fortunately it doesn't. As the pure-bred ewes get older, they are less able to cope with the rigours of foraging on the hill and they are 'drafted'; that is, they are sold away, generally as 4-crop ewes, to farmers on rather kinder land, where they can often produce another two or three crops of lambs.

This, then, is our first move down the hill to what we may call the Uplands.

The Uplands

I've said 'rather kinder land', and certainly there are less severe climate conditions, but we are still talking about marginal land. All things are relative, but they are the conditions where the draft hill ewe can not only produce lambs for several more years, but also bigger and better lambs. The essential element is that she is able to do just this – that she can be put to good and profitable use in her latter years, when otherwise she would have to be culled.

So we come to cross-breeding, and the ram traditionally used for this purpose is the so-called 'Longwool' ram – breeds such as the Border Leicester and the Bluefaced Leicester. The purpose of the exercise is not only to produce bigger lambs, growing faster to bigger weights and thus increasing the output from the land, but also to produce a half-bred ewe lamb suitable for use as a fat lamb mother under fertile lowland conditions. Put like that, it sounds as if the whole thing had been worked out beforehand but, of course, it wasn't like that at all. This link in the chain just evolved as breeders found that the half-bred ewe produced by this cross was a very valuable addition to their sales output.

The Lowlands

Here – 'downhill' – we have the richer, more fertile, and therefore much more costly conditions of the lowlands, be they in the grassland West or the arable East. The high cost imposes the discipline of high production and the sheep flock has to justify itself in competition with other high output alternatives such as dairying or cereal production. A very different role indeed from that of the pure hill ewe where no such alternatives exist.

The lowland ewe then has to be productive not only individually, but also per hectare in that she must be capable of accepting high stocking rates, a factor which imposes the need for a particular type of hardiness. She must be prolific and a good milker and she must live a long time, but it is generally accepted that she does not herself have to be the best carcase sheep in the world. The reason for this is simple: it has been found to be difficult to the point of being impractical to combine in one animal the natural qualities of prolificacy and milk yield with those of carcase quality.

So we come to the last link in the chain – the so-called 'Down' ram. Bred and selected entirely for their carcase qualities, the Down breeds, notably the Suffolk, are crossed with the half-bred ewe to produce the final product, the butcher's lamb.

THE ADVANTAGES OF STRATIFICATION

There are many advantages and not least among these is the fact that nationally we are making good use of the many different land and climatic conditions that we have in the United Kingdom. Good could always be better, of course, but we can perhaps best recognise our good fortune when we look at a country like France, whose hill sheep farming has been allowed to deteriorate to the great disadvantage of her whole sheep industry. In a later chapter I shall be looking at this in more detail.

But standing out above all other considerations is the fact that taking the pure hill breeds, the pure 'Longwool' breeds and the pure 'Down' breeds and putting them together in a systematic and organised chain, we are not only making use of the qualities of those pure breeds themselves, but we are also cashing in on the value of the phenomenon of 'hybrid vigour'. To this precious asset we can add the benefit of moving sheep downhill, from hard conditions to a softer life so that at each stage the sheep improve.

It is such a logical progression that one is tempted to say that it could only have happened by accident!

THE POINTS OF ATTACK

However, we must not be complacent. The overall productivity of our national sheep flock is pretty dismal, and whilst much of this is due to management, there is nevertheless much to do in breeding and selection. Our system of stratification gives us two obvious points of 'attack'.

First of all, we ought to recognise that there is little we can do to improve the hill breeds. Some of the breeds themselves are better than others, but within each breed their qualities are determined principally by exposure to their environment. In any case, flock productivity is such that selection is virtually impossible – nearly all the ewe lambs have to be kept as flock replacement. Improvement can therefore only come via the pure-bred hill ram and the scope is very limited. However, when we come to the category of 'Longwool' ram, it is another story altogether. The numbers used are relatively small, but their influence is enormous. The 1972 MLC sheep survey showed that 1·15 million hill ewes were put to Longwool rams out of a total of 5·5 million of all hill breeds. This survey has unfortunately not been repeated but we know that hill breed numbers have increased to over eight million and of these some two million are put to Longwool rams. The production of these rams is concentrated into small flocks of very few breeds. Clearly this is a point where we have a great chance to influence the performance of the half-bred ewe, provided always that our objectives for selection are clear and logical, a point I shall return to later.

The second, and very important point of attack, is the Down ram. Obviously, half the genetic make-up of the final product in the chain comes from these breeds. And indeed often more than half, as the Down breeds are frequently used outside of their classical place in the structure – either as a direct cross onto a hill ewe instead of a Longwool ram or as a second cross, *e.g.*, a Dorset Down ram onto a ewe which is a Suffolk Half-bred [i.e., Suffolk x (Border Leicester x Cheviot)]. Here, as with the Longwool, the influence of the Down breeds is out of all proportion to their numbers and there is great scope for improvement – which takes us on to breeds and breeding.

Chapter 3

BREEDS

IT IS NOT easy to write about sheep breeds, and even less easy to write objectively. Everyone has his own favourite breed, and there are countless favourites.

No fewer than 50 pure breeds were recorded in the 1972 MLC survey, but only (!) 24 of these had more than 25,000 ewes. Over 300 specific crossbred types were named, but again only 30 were numerically important. Tables 2 and 3 and 4* give the details – and these not only confirm the overall pattern of stratification but also illustrate the extent to which it is fragmented.

Lest anyone should be getting alarmed at the prospect of me going through the list breed by breed, let me reassure them. That would require a book to itself. An admirable example is *British Sheep* (National Sheep Association). Instead I propose to think aloud, as it were, about the main breed categories and try and sort out just what it is we ought to be doing about them.

THE HILL BREEDS

It is easy to understand how we have got where we are with hill sheep. They have evolved from local types of sheep, cut off from each other by non-existent communications until the last 100 years or so – each breed type adapting itself to survival and production under what are often miserably difficult conditions. For instance, some breeds, like the Herdwick, have an astonishing ability to survive on the rain-swept slate of the Lake District, but perhaps have little other economic merit to commend them. Others much more numerous, are remarkable not only for their hardiness but also for their productivity.

* Source: MLC Sheep Breed Survey 1972.

Table 2. Pure Breeds occurring in the MLC Sheep Breed Survey grouped by Breed Class

HILL BREEDS	Scottish Blackface Swaledale Dalesbred Derbyshire Gritstone Lonk Herdwick Rough Fell North Country Cheviot South Country Cheviot	Welsh Mountain Improved Welsh South Welsh Mountain Black Welsh Mountain Radnor Beulah Speckleface Hardy Speckleface Exmoor Horn Shetland
UPLAND BREEDS	Clun Forest Kerry Hill Devon Closewool	Greyface Dartmoor Whiteface Dartmoor
ROMNEY	Kent or Romney Marsh	
DEVON	South Devon	Devon Longwool
LOWLAND	Dorset Horn Polled Dorset Horn Improved Dartmoor Jacob	Wiltshire Horn Lleyn Llanwenog
LONGWOOL	Border Leicester Bluefaced (Hexham) Leicester Teeswater Wensleydale	Leicester Colbred Cambridge Finnish Landrace
DOWN	Suffolk Dorset Down Hampshire Oxford Down	Southdown Shropshire Ile de France Ryeland
NEW DOWN BREEDS	Texel Meatlinc	Charollais

Table 3. Numbers of Ewes (including Ewe Lambs) put to the Ram in Great Britain (excluding Shetland) in Autumn 1971.
(Grouped by Breed Class of Ewe and by Breed Class of Ram to which the Ewes were put).

Ewe breed class	*Ram breed class*	*No of ewes (thousands)*	
Hill	Hill	4,601	
	Longwood	1,151	
	Upland	76	
	Down	979	
	Others	89	6,896
Upland	Upland	252	
	Down	506	
	Others	59	816
Romney	Romney	150	
	Down	131	
	Others	13	294
Devon	Devon	50	
	Down	109	
	Others	19	178
Lowland	Lowland	51	
	Down	31	
	Others	9	91
Longwool	Longwool	14	
	Down	10	24
Down	Down	232	
	Others	11	243
Longwool x Hill	Down	1,808	
	Others	135	1,941
Upland x Hill	Down	219	
	Others	81	300
Down x Hill	Down	119	
	Others	35	154
Down x (Longwool x Hill)	Down	597	
	Others	35	632
Other Cross-breeds	Upland	66	
	Down	396	
	Others	41	502
GRAND TOTAL (thousands)			12,072
UPDATED GRAND TOTAL 1983 (thousands)			15,500

Table 4. Numbers of Ewes of Hill Breeds (including Ewe Lambs) put to the Ram in Great Britain (excluding Shetland) in Autumn 1971.

(Classified by Breed of Ewe and Breed or Breed Class of Ram to which the Ewes were put).

Ewe breed	*Ram breed*	*No of ewes (thousands)*	
Scottish Blackface	Scottish Blackface	1,533	
	Other hill breeds	136	
	Longwool	563	
	Down	98	
	Others	8	2,338
Swaledale	Swaledale	272	
	Other hill breeds	15	
	Longwool	185	
	Down	32	505
Dalesbred	Dalesbred	126	
	Other hill breeds	16	
	Longwool	47	
	Down	7	196
Welsh Mountain	Welsh Mountain	1,157	
	Other hill breeds	174	
	Longwool	140	
	Down	413	
	Others	82	1,966
Beulah Speckleface	Beulah Speckleface	95	
	Longwool	2	
	Down	95	
	Others	13	206
Hardy Speckleface	Hardy Speckleface	158	
	Other hill breeds	9	
	Longwool	3	
	Down	126	
	Others	32	327
North Country Cheviot	North Country Cheviot	311	
	Other hill breeds	5	
	Longwool	128	
	Down	60	
	Others	1	504
South Country Cheviot	South Country Cheviot	228	
	Other hill breeds	6	
	Longwool	24	
	Down	43	301
Other hill breeds	Other hill breeds	362	
	Longwool	59	
	Down	103	
	Others	30	552
TOTAL HILL SHEEP			6,896

Not only have overall numbers changed between 1971 and 1983 but so also have the proportions of one breed to another. The increases have taken place mainly in Wales and the Pennines – thus Swaledale and Dalesbred have gone up by 31% and the Welsh breeds by no less than 39%.

I have to admit that I find it difficult to restrain my admiration for the Scottish Blackface – and that will gain me no friends at all amongst breeders of the Swaledale, the Cheviot, and the Welsh Mountain sheep. I write, of course, as an ignorant lowlander and my opinions can therefore be dismissed by the 'cognoscenti' who really do know what they are talking about. Nevertheless, I cannot be considered a detached observer from afar. I am, after all, a customer for the product in the form of the half-bred ewe, and it is quite simply my experience that the 'Blackie' is among the best of parents. She has, however, one distinct disadvantage – she has horns. Not that this matters with the Greyface ewe – the horns rarely come through to the half-bred – but it matters desperately to those who purchase and fatten hill store lambs. Blackface wether lambs and lowland fencing, particularly electric netting, just do not go together.

I have to say that these horned lambs have to be cheap indeed for me to want to buy them, as opposed to their hornless competitors. The Scottish Blackface breeders could render no greater service to their breed than to breed off these useless but attractive appendages: not an impossible task but admittedly a difficult one.

Horns apart, we must consider the fundamental question – should anything be done to improve our hill breeds, which are, after all, half of our national flock? So much of the potential improvement in our hill flocks lies with better management, particularly of the hill grazings, which in turn depends on the confidence coming from cash income to generate the large sums of capital required. So much depends on management that it is not easy to see just where the potential lies for genetic improvement.

Individual ewe recording, which must be the basis for any serious selection programme, is almost impossibly difficult on the hill. Take the ewes off the hill in order to record them and you take them away from their environment, which imposes its own special and essential selection.

Do we want to improve the hill ewe? If I say that, personally, I doubt it, I am perhaps avoiding the issue because of the difficulties involved; and to suggest that no improvement is possible would be wrong indeed. Substitution of one breed with another will go on as it is now; for instance, with the Swaledale replacing the Herdwick. Within any one breed, there will always be breeders who are more gifted in the practice of their art than others. But fundamental improvements from nationally-supported schemes of recording and selection, that I very much doubt. However, I have to say that I have been very impressed by what I have heard recently of Group Breeding Schemes. Reference to these is made later in the chapter on Breeding, and I begin to believe that such schemes could have considerable merit with Hill Breeds. Nothing is so good that it cannot be improved further.

THE LONGWOOL BREEDS

One of my key points of 'attack', for here we really could have some influence on the profitability of lowland sheep production (second only to what we could achieve with the Downs breeds). The influence of the 'Longwool' crossing ram hardly needs underlining. Taking the MLC 1972 survey once more, we can see that no fewer than 1·28 million cross-bred ewes are sired by the Border Leicester alone. By 1983, we know that the figure for cross-bred ewes sired by the Border and Bluefaced Leicesters together amounts to well over 2 million. Only the Suffolk, in the Down category, has greater influence. This being so, there is clearly a case for putting some nationally supported effort into the breeding and selection of these rams.

What are we seeking to achieve? All the time we must keep firmly in our minds the sole objective – the half-bred ewe. On her performance the Longwool ram must be judged and *not* on what he and his breed look like. So the genetic qualities in the ram must complement those in the hill ewe. She will certainly be hardy, a good forager, and a good milker when fed properly. But she will also be only moderately prolific, small and a slow grower, and necessarily have that nervous, alert disposition which is part of her survival mechanism on the hill.

Ram Selection

Briefly, and I shall return to this in greater detail, the characteristics we should be selecting for in the ram are:

(a) prolificacy;
(b) calmness and docility of temperament;
(c) fast growth rate;
(d) longevity.

These must be our main aims, on which we must really put pressure. There are, of course, other objectives which we cannot neglect – wool, and where on the body that wool is to be found; physical soundness; and a degree of bodily conformation – but these must remain secondary.

Table 5. Statistics of Some Registered Sheep Flocks

Breed	*No. of flocks*	*Av. No. of years reg'd*	*Av. No. breeding ewes*	*Av. No. of rams per flock*	*% of home bred rams*
Bluefaced Leicester	350	19*	10	1·4	NA
Border Leicester	397	NA	24	2·3	9
Clun Forest	320	NA	129	3·1	14
Devon Closewool	190	18	191	3·5	8
Dorset Down	77	14	103	3·0	16
Dorset Horn	107	15	87	2·9	21
Hampshire Down	66	15	59	2·2	24
Leicester	43	29	32	NA	NA
Lincoln	20	20	43	1·4	14
Oxford Down	32	30	34	1·6	8
Polled Dorset Horn	37	6	94	3·4	37
Romney	57	17	271	8·1	53
South Devon	93	20	53	1·8	0
Southdown	25	NA	64	2·5	32
Suffolk	915	12	40	2·2	13
Teeswater	195	22*	6	1·7	NA
Wensleydale	17	22*	10	1·3	NA

NA = not available.
* *Average age of flocks (flock book not in existence for some of the period).*
Source: Data drawn from breed society flock books.

Of the breeds in current use, two stand out way above any of the others: the Border Leicester, with the Bluefaced (Hexham) Leicester coming up rapidly to challenge for first place. The vast majority of our Halfbred ewes, be they the stylish Scottish Halfbred out of the Cheviot, the Welsh Halfbred, or the various Greyfaces, Mules and Mashams, are sired by one or other of these ram breeds.

These two breeds have stood both the test of time and that of acceptability by the lowland customer. And there are many who would rest satisfied with that and say 'why look further?' Yet no-one can say that the average lowland flock owner is doing his job anything like as well as he should and making as much money as he would wish. Judged by the standards of the Halfbred ewes' performance – and even admitting that this is very much affected by sub-standard management – there is a need to produce a much better ewe than anything we have at present. The cold facts are set out in that admirable publication, the *MLC Sheep Yearbook*. In 1983, the average lowland spring lambing flock sold 1·46 lambs per ewe tupped at a stocking rate of 13·1 ewes/ha. Even with the high lamb prices of the 1980s, that sort of performance just does not compete with alternative forms of land use. Even the top third of MLC recorded producers do no better than 1·49 lambs reared at a ewe stocking rate of 16 per ha.

Faced with these facts, one would have expected to see improvement work going on amongst the Leicester breeders. Yet this is not the case and it is easy to see why. The average size of registered flocks of Border Leicesters is only 24 ewes (397 flocks) and that of the Bluefaced Leicester a mere 10 (350 flocks) – Table 5. Overall numbers of each breed are really very low and flock size is tiny. Hardly any breeder has sufficient ewes to enable him to be self-contained and use rams of his own breeding – which, to my mind, is fundamental to making real progress.

Instead there is an interchange of rams between breeders and choice is made on the basis of 'breed points' and not on any recorded facts. So we get the ludicrous situation where breeders get more excited about the exact shape of the ram's Roman nose or the arch of his back than about the performance of his daughters; about which they neither know nor care. No doubt, this is fun for all concerned but it does nothing whatsoever for the profitability of my lowland flock.

It is just this combination of tragi-comic circumstances which has stimulated a few individual breeders as well as the Animal Breeding Research Organisation (ABRO) to do something about it. It would be pleasing to be able to record a success story and to say that here we have a breed which is coming up to challenge the dominance of the two Leicesters. But that is not the case, though perhaps it could be said that there are one or two waiting in the wings. It is a surprisingly poor reward for what has been a very great deal of effort all concentrated in the last 25 years.

The Colbred

There was, first of all, the Colbred: the first work of constructive breed creation that had been carried out for over a hundred years. Oscar Colburn, of Northleach in Gloucestershire, saw the opportunity and the need for a new and better Longwool ram. The result was the Colbred: a combination of four breeds, the Border Leicester, the Clun Forest, the Dorset Horn and the East Friesland milk sheep. As a breeder, Colburn was undoubtedly successful, for the Colbred ram produced half-bred ewes of considerable merit, of higher prolificacy and of earlier maturity.

The Colbred looked like establishing itself as a breed to be reckoned with. Where Colburn failed was in commercialising his work. For he sold out the major part of his interest to Thornbers, who sought to repeat, in the sheep world, their success with poultry. This turned out to be a disastrous mistake, as the poultry agri-business men tried to go too far too fast without much regard for either stockmanship or the cautious conservatism of their potential customers. Thornbers soon pulled out, and the Colbred remains in small numbers with some few enthusiasts, and it may be that it will come again. The sad thing about this episode is that, not only did it bring about the downfall of the Colbred, but it also brought the work of the geneticists as applied to sheep into disrepute.

The Finnish Landrace

At around the same time, in the 50s, the main preoccupation of research was with prolificacy. The major overhead cost of lamb production, it was argued, was the cost of maintaining the ewe for the whole year; and at best, at the end of it, she produced not much more than a lamb and a half. If only we could put that figure to somewhere between two and three! So the Finnish Landrace was imported, one of two breeds in the world with remarkable prolificacy (the Romanov being the

1. Border Leicester ram.

2. Bluefaced Leicester ram.

3. Cambridge ram.
Source: Cambridge Sheep Society.

4. Scottish Blackface ewes . . . typical hill breed used to produce Halfbred ewes by a Longwool ram.
Source: Author.

other). The ability to produce litters of three and four lambs is certainly there, but, apart from that, the Landrace is a pretty unattractive sheep: conformation, size and growth rate all being quite unacceptable. To be fair, even the most enthusiastic recognised this and saw the role of the Finnish Landrace only as part of a mixture with something more attractive.

The 'Cadzow Improver' and 'Dam Line'

Out of this importation have come two 'breeds'. Firstly, the so-called 'Cadzow Improver' produced by Brian Cadzow, the make-up of which was never revealed but which was understood to be a Dorset Horn/Finnish Landrace cross. Cadzow repeated Colburn's mistake, went commercial with a succession of industrial companies and failed. Once again, and quite unfairly, the reputation of the geneticists took a hammering.

Then secondly, there has been the work of ABRO in producing the rather badly-named 'Dam Line'. Once more it was a mixture of different breeds – the Finnish Landrace, the East Friesland, the Border Leicester and the Dorset Horn – with the same objective. This work is well within the terms of reference of ABRO, who have a duty to push out the frontiers of breeding in a way which is often not possible for the individual breeder who has to earn his living in the short term.

With the financial resources of the Agricultural Research Council behind them, the scientists at ABRO were able to ride out the difficulties which obliged both Colburn and Cadzow to sell out to the speculative industrial companies. The end-product, the 'Dam Line' breed, has a great deal to commend it as a sire of a Half-bred ewe. Its performance is well documented and all the figures show that it has sufficient virtue to become a serious competitor to the Leicesters. The question is: can it be carried on into sufficient numbers so that it can become of significant importance? For, at the moment, it is 'stuck' at ABRO with no obvious route into commercial exploitation.

The Cambridge

Finally there is the Cambridge, probably the most prolific of all British breeds. The story begins in the early 1960s when Professor John Owen, now at the University College of North

Wales at Bangor, was lecturer in Agriculture at Cambridge – hence the breed name. He and his colleague, Alun Davies, now at Liverpool, were of the view that the way to higher profits with sheep lay with increased prolificacy. Unlike many others at the time, however, they did not believe that it was necessary to go outside British breeds to get highly prolific stock. They recognised that, within our own breeds, there were outstanding individuals whose performance was consistently way above average. Was it not possible to collect together a flock of these individuals to form the basis of a new breed?

This is exactly what John Owen and Alun Davies did. They advertised in the agricultural press for ewes that were proven triplet producers. It did not really matter what colour or shape they were, so long as they had the essential quality of prolificacy. In fact most of them came from the Clun, Llanwenog and Llyn breeds. One can just imagine that they looked a pretty mongrel lot once they were all assembled together! The next stage was to use Finnish Landrace rams, the only imported bloodstock, as a once and for all cross on these ewes to produce the F1 generation. This was followed by twenty years of rigorous culling and selection based upon performance records until now in the mid 1980s the Cambridge is an established pure breed. That is quickly and easily stated – the reality of course was made up of many years of hard work as well as faith in the achievement of the final objective. It couldn't have been done without the co-operation between the universities and commercial farmers. Both John Owen and Alun Davies would be the first to pay tribute to that willing co-operation – nevertheless much of the credit goes to them for the inspiration in the first place, and the leadership and discipline subsequently necessary to turn a good idea into a working Breed Society.

The question has to be asked – is the Cambridge anything more than an interesting academic exercise? Undoubtedly it has been shown that it is possible to select out of low performance populations, another population of very different performance characteristics. There is a truth here which ought to be far more widely appreciated than it is – that our sheep breeds are no more than populations which basically *look* alike. Behind the visual similarity, there lies a very wide divergence of characters all of them ranging from the very good to the very bad. Progress in breeding therefore depends upon identifying the very good and eliminating the very bad – taking off the bottom 50 per cent, or better still the bottom 75 per cent but that is luxury indeed, and culling them.

The Cambridge, then, has proved that this works, for it certainly is a very prolific breed indeed. The average performance for mature ewes, under good management, is over 300 per cent lambs born alive. Prolificacy however has not been the sole criterion for selection. Early maturity is considered to be of great importance and ewe lambs have to breed at a year old if they are to be considered for retention in the pure flock. Milk yield as measured by lamb growth rate is the third major characteristic which is recorded and put into the index on which selection is based. The result is a pure breed of acceptable standards of uniformity which, under the good management which is essential if it is to express its potential, can perform at very high levels of output.

So the answer to my question is that the production of the Cambridge has most certainly been an interesting academic exercise, but at the same time one with considerable potential for practical application. The place for the Cambridge must surely be as a competitor for the Leicesters – to sire a half-bred ewe that is capable of producing a consistent 200 per cent lambing percentage. Though that has yet to be proved, what is undeniable is that it is an objective worth pursuing. For we seem, with the Leicester half-breds to be stuck on a performance plateau of around 175 per cent and under present conditions of cost and return, we must at least be striving to go further.

At Worlaby, we have started our own trials to evaluate the Cambridge cross, in our case with the Scotch Blackface. I want to see if we can push our output up from the ewe, whilst at the same time maintaining our high stocking rates. The first ewe lambs lambed down in April 1984 and we weaned 148 per cent from them, which from ewe lambs is good going indeed, and moreover virtually all of them were in lamb. We have to wait and see what they will do as mature ewes. It is certainly going to be at a higher level than anything we have used before and one of the questions in my mind is whether we have the management ability to cope with much higher numbers of lambs born. But that is a challenge to me!

The Future of the Longwool Breeds

It almost goes without saying that this collection of breeds is a vital link in the structure of our sheep industry. The only question is which breed can serve us best. I have already drawn attention to the potential for improvement by selection based upon recording which exists in all the breeds. And nowhere is this more the case than with the Border and the Bluefaced Leicester. Professor Owen has shown what can be done by creating the Cambridge from amongst a motley bunch of individuals. How much easier it ought to be within a pure breed which is already established in public favour. Sadly we are up against the barriers of pedigree tradition.

So will the new Longwool breeds, the Cambridge or the Colbred, or even the ABRO Dam Line become serious challengers? I would like to think so if only because competition is good for all concerned. But whether they will or not will depend not only on their individual merit, but also on the ability of their breeders to overcome a significant obstacle. It takes time, a lot of time, and therefore a lot of money, both to create a new breed and also to bring it to the point of commercial usage via its half-bred offspring. There is a basic dilemma here, a dilemma which Colburn and Cadzow tried to get around by teaming up with the resources of industrial companies. They failed because such companies have not got the patience to wait – they can only succeed with the rapid multipliers – the pigs and the poultry. Cattle and sheep breeding must remain with the breeder, but perhaps we should consider that the breeder should lose some of his individuality by working on a much larger scale under a Group Breeding Scheme. This may be the only way, under the rigorous economic conditions of the twentieth century to encourage the flair and initiative of the individual breeder artist and to transform his work into a scale of national significance. Robert Bakewell, the great improver of the Leicester sheep in the 18th century, could not have operated on his own in today's conditions.

THE DOWN BREEDS

This is the second key point of 'attack'. The sole purpose in using the Down ram is so that, when crossed with the half-bred ewe, he puts carcase quality into the subsequent butchers' lamb. Put starkly like that, it sounds simple; no complications trying to select for a combination of characteristics as with the Longwool breeds. Just a single-minded concentration on meat. But it isn't simple at all – it is, in fact, extremely difficult, and for one basic reason. We have not yet found the means of 'measuring' meat quality in the live animal, so selection in the meat breeds depends upon 'judgment'.

How has this judgment operated and what has it led us to? As the Longwool breeds are dominated by the Leicesters, so the Down breeds are dominated by the Suffolk. In 1972, over one-third of the national ewe flock, over four million ewes, were mated to Suffolk rams; and a visit to any fatstock auction in the country will confirm the popularity of the black-headed lamb. All this would seem to lead to the conclusion that the judgment exercised by Suffolk breeders in response to their customers requirements has been sound.

Yet the facts are very different. Go to any big abattoir, particularly one with an interest in the export trade, and you will hear one recurrent complaint. That no matter how skilled the buying and whether that buying is done direct from farms or from an auction, it is impossible to buy any run of lambs of a consistent and uniform quality. Some part of the reason for that is, of course, due to bad management and selection on the farm. But equally a very significant part of the blame must be laid at the door of the genetic quality of the ram. Although many of these lambs will be Suffolk crosses, there will also be many that are the product of other Down breeds such as the Dorset Down or the Hampshire Down. There seems to be little to choose between them as regards consistency of carcase quality. More serious still, if one goes to one particular flock where presumably the selection criteria are evenly applied, one still finds this lack of consistency in carcase quality. I am generalising, of course, but at least this has been my experience; and the grumbles from the big abattoir owners are so unanimous, that I believe that this criticism must be justified.

And if it is spread across all breeds, then it must be our methods of selection, of 'judgment', that are at fault.

There is therefore a very big job of work to be done here; and a very worthwhile job if only we can find the right way to do it. Whether that way will turn out to be an improvement of what we have already got or whether it will come from those new breeds now coming forward, remains to be proved.

The Texel

Of the 'new breeds' in the Down category, the Texel has become a serious contender. It is not, of course, a new breed at all, but an import from France (although it is really a Dutch breed originating from the island of the same name). The Texel has a well documented reputation for producing extra lean carcases, thus conforming to modern market requirements both on the home market and for export. It has however the disadvantage of being a relatively slow grower, which makes it more suitable for the production of autumn and winter lamb.

The Texel has been in the country long enough and is established in sufficient numbers – there are said to be some 7,000 registered females – for us to make some sort of judgment as to its future. It got off to a great start with the benefit of widespread publicity organised by the original importers. All the talk was of the Texel and how it was going to transform the quality of British lamb. Very high prices were realised at auction for individual rams and ewes – and those of us with long memories were reminded of the similar circumstances of the first importation of Landrace pigs. History of course repeated itself and the consequence was the same – half-breds, quarter-breds, even those with one-sixteenth Texel blood were kept and bred from and were called Texels. Inevitably quality suffered and disillusion set in. Texels were too fashionable, and thus too expensive and of too uncertain a quality to be of interest to the commercial lamb producer.

To be fair, that is now behind us and the Texel, like the Landrace pig, has got over the disability of its original extravagant promotion. It is established, it is a breed with considerable qualities and potential for a significant contribution to the British sheep industry. It is also fair to say that it has not made the progress that was within its grasp. Many Texels

5. Suffolk ram.
Source: Farmers Weekly.

6. Texel ram.
Source: Douglas Low.

are far too short and disappoint when the weights come back from the abattoir. The reason for this lack of progress is not hard to find and is, frankly, particularly saddening to anyone who has the long-term interests of the industry at heart. For the Texel breeders have created a pedigree structure very similar in many respects to the Leicester Breed Societies. Many small flocks, very few of them indeed large enough to breed their own ram replacements. Dependent therefore on buying other breeders' rams chosen more often than not on a visual appreciation of beauty rather than the cold facts of performance records. It is all very sad. Remember that any breed is but a collection of animals that look alike and progress can be made either upwards or downwards, depending on how selection is made.

The Charollais

Another imported breed, again coming from France, where it is one of the numerically smaller breeds. An important one though where it is used, like the Vendean and the Charmoise for the production of autumn-born lamb finished at Easter. A type of production where the sheep are kept indoors the whole time. Its best friends in France would admit that as a breed it is very variable – and that it is very important to select a good Charollais. And it must be said that when it is good, it can be very good. I have used the blood, albeit sparingly, in the make-up of the Meatlinc – and I did so because at its best, the hindquarter development can really be very impressive. But as the good can be very good, so can the bad be horrible! Nevertheless there is genetic material there of considerable worth to be made use of.

However, the Charollais in its pure form has one great disadvantage for British conditions. Its lambs are born virtually naked, and this absence of birthcoat is a serious disability under British conditions. It does not matter a hoot in France, of course, because the lambs are always indoors. And sadly this trait continues to show in its half-bred progeny so that crossing the Charollais ram with a woolly half-bred ewe still tends to produce lambs which are very vulnerable to the British climate in their first two or three weeks of life. My conclusion is that the place of the Charollais is in a breed improvement programme and certainly not as a pure-bred terminal sire.

THE MEATLINC

The third contender of the 'new' breeds in the Down category is my own Meatlinc. I may perhaps be forgiven, if in my own book, I devote some considerable attention to this newcomer to the sheep breeding scene.

How it All Started

The Meatlinc was born out of frustration, way back in 1964. I was, in those days, buying Suffolk rams to use on our flock of what was then mostly Clun Forest ewes. I went to some considerable pains to buy good, and certainly expensive, Suffolks from breeders who had gained a reputation country-wide. For I had always believed, as I do now, that to skimp on the cost of a sire, be it ram or bull, is false economy. Yet I was never really satisfied with what I bought. For one thing, I was not able to find a breeder who could *tell* me anything about his sheep. Apart, that is, from whether they had come from such and such a prize-winning strain; and in some cases, whether they were twins or singles. Secondly, I never felt happy with the rams after I had had them on the farm for some while. Long enough for the sales preparation bloom (*i.e.,* overfeeding and trimming) to have worn off. By then it was obvious that there was a lack of uniformity, and worse still, that much of what appeared to be meat on the hindquarters and the loin, was not meat at all, but fat. The subsequent lamb crop sired by these rams showed all these faults. A wide variation in type with nothing like enough of them showing real 'meaty' characteristics.

As I say, I felt frustrated. Perhaps I could have looked further and found better, but I doubt it. Instead I decided to do something about it myself. The question was, what? Should I take a British breed, well known and accepted like the Suffolk, buy a nucleus of outstanding individuals so as to form a small flock, set up the appropriate selection criteria and go on from there?

Alternatively, should I go out on a limb, take a leaf out of the Charolais cattle breeders' book, and import a nucleus of a chosen pure meat breed from abroad, say, France. My visits to France at around this time had indicated to me that some of their meat breeds such as the Ile de France, the Berrichon du

7. A flock of Greyface ewes (Border Leicester × Scottish Blackface).
Source: Topham.

8. Meatlinc ram.
Source: Author.

Cher and the Vendean, had outstanding carcase qualities. Not that French breeders were any more gifted than their British counterparts, or any less mesmerised by winning prizes at, in their case, the Agricultural Salon in Paris each March. None of these things – but what did influence commercial French breeders was the fact that their market paid a very high price for high quality and very little indeed for poor quality.

The Choice

I looked seriously at both these alternatives and finally discarded them. There were a number of reasons for this, but none of them included the fundamental one that success was not possible with either. I am certain that had we applied exactly the same methods which we have followed in the selection and production of the Meatlinc, we would have made worthwhile progress with any chosen pure breed in the Down category.

Why, then, did I go for the third alternative, which was to select superior individuals from a number of different meat breeds, put them together, select from them and eventually produce a new breed? I think I must admit that probably the most important reason was that if we were going to put a great deal of effort into it over a long period of years, then I wanted to produce something distinctive at the end of it all. Not just another flock of, say, Suffolks – one amongst many, better than some, no doubt not as good as others. There were other, and rather less pretentious reasons. I really did think that some of the French breeds had got something to offer. The quality of fleshing, particularly on the hind leg, was impressive, as was the absence of fat overall in the carcase. On the other hand, most of them were bred and selected indoors, often lambing out of season, and therefore being tested under totally different conditions from our own. I had no means of knowing whether, in pure form, they would adapt to our more rugged, grass based systems. There seemed therefore a great deal to be said for using a base flock of British ewes onto which we would introduce imported French blood.

How it Was Done

This, then, was exactly what we did. The base flock of ewes

selected as the starting point was a mixture – the common features were size, for I was convinced that high adult weight was positively associated with fast liveweight gain; and conformation, that is to say, carrying plenty of flesh down the hind leg and across the loin, and doing this under plain, commercial conditions of grass feeding. Breed was of no importance, and I could not have been less interested in what colour the head was. On to this female base flock, we put individual rams of different breeds, chosen not because of the reputation of their respective breed, but because I thought that as individuals, they had the qualities I was looking for. Over the course of a number of years, individuals from the Suffolk, Dorset Down, Ile de France, Berrichon du Cher, and Charollais breeds were used. In addition, we had a short flirtation with the Texel, but quickly discarded it because it slowed down growth rate and visibly shortened our sheep.

This sounds like a glorious and totally illogical mess of pottage! Well, it was certainly a mixture, but equally certainly, it was not a mess. We had put into a mixing bowl, size and growth rate plus high carcase quality. Not all in the same animal, male or female, Not all to the same degree. Out of the mixture, we had to extract those combinations which, like cream, came to the top and which met my idea of what we were aiming for.

To do this we divided the flock into five different female families, each containing around 60 ewes, to give a total of over 300. This size is important. To come much lower in numbers leads to the risk of running out of genetic manoeuvring room. This incidentally, is the problem with very many pure-bred flocks – they are just not big enough to select from within and they are forced to go outside to buy rams. And as soon as you do that, you buy the unknown, and you can go backwards as fast as your go forwards. It was a fundamental principle with me that we should have a closed flock, selecting from within; once, that is, we had put into the mixing bowl the best that could be found.

The five families then were set up and they were recorded under the MLC individual ewe recording scheme. Now records can be masses of figures on mountains of sheets of computer paper, or they can be the basic information, the bricks with which you build your edifice. Recording is quite simply the means of identifying those animals which are superior and on which you wish to base your future breeding. The first priority for us was to sort out and eliminate, out of the mixing bowl, those individuals that were the poor performers. This brought about a dramatic and rapid change. It is the first stage in the improvement of any population. It is the first stage – it is also the easy one. The second stage – that of selecting steadily better and better individuals from within the population and spreading their influence – takes much, much longer. The better you get, the slower becomes the progress.

One of the ways by which it is possible to speed up change is by going for a rapid generation turnover – by always using ram lambs and by keeping the average age of the ewe population low. It is really essential to do this, or a lifetime will not be long enough to get anywhere. As it was, we started in 1964 and it was twelve years before we had a saleable product in the form of shearling rams.

For many practical reasons, I decided to put the weight of selection pressure onto the male side. We just could not afford to select the females too hard or we should have run out of numbers. Our procedure for doing this involved an assessment each August of that year's crop of lambs. Ram and ewe lambs from each family are penned separately. For me, it is one of the most interesting days of the year. To have a good look at the different families, as distinct from the total population. Then to go on and pick each family to pieces, as it were, and sort out the high performers and the good lookers.

Order of Attack

We have always followed the same order of attack. First of all, we go through the five pens of ram lambs. Using the MLC records, we throw out all those individuals whose performance is not good enough. In the process, we find that we often throw out some individuals that on looks alone we would have certainly kept. It needed considerable self-discipline, especially in the early days, to disregard the pleas of my shepherds to keep certain rams that they particularly liked the look of. But one must be hard! We are then left with a reduced number of the better performers. It is only at this stage that we actually look at the lambs. Are they good enough physically? And is their meat conformation what we are looking for? So a few more get rejected. Then out of this remaining population, all of which will be kept on for eventual sale as shearling rams (although in the intervening time, they are subject to at least three more selection checks), we have to select those few outstanding individuals who will be used for stud breeding that season. Generally, we bring the number down to three, based both on performance and on looks.

We follow the same procedure with the ewe lambs, except that we do not have to select the top few. It is on the female side where we are looking for uniformity as well as quality. At the same time as we are looking at the ewe lambs, we also have a close look at least year's crop, the gimmers, to see how they have grown out since we subjected them to this selection a year previously. Inevitably, some will have disappointed us, and must go. The same is true of the older ewes, and as I said earlier, where you are seeking rapid improvement, then you just must keep the flock young. That is not to say that you should not keep the outstanding 'matrons' – of course one should, but they should be subjected to very close scrutiny indeed.

The breeding flock is thus set up for the following year. What I have not explained is how we maintain the circulation of improvement and at the same time avoid getting too close to in-breeding. I have said that there are five families, and let us number them 1 to 5. The ram lambs finally chosen for stud use from Family No. 1 are used on Family No. 2. Those from No. 2 onto Family No. 3, and so on. The females always remain in their original family. The males are always crossed onto the next family in line. The other precaution is that once any individual ewe has produced a ram lamb that is eventually selected for stud use, that ewe is never used as a stud mother again, She is kept in the flock, of course, because she is obviously an outstanding individual and she will breed ewe lambs that can be kept, as well as ram lambs that can be sold.

The Basis for Selection

Throughout this brief description, I have often used words

like 'performance' and 'conformation'. What exactly do they mean? Or are they just a con-trick, a cover-up to try and conceal that I was no more than choosing what I liked the look of.

To answer that question, we have to go back to what we are trying to achieve. And this is the definition of the Meatlinc.

(1) It must be a breed which, when crossed with commercial half-bred ewes, is dominant – it must stamp its authority and its quality on its progeny.
(2) It must be quick growing, and I am sure that this means that it must be big.
(3) The bigness must, however, be combined with quality. If you select on growth rate figures alone, there is a real danger that you will finish up with big, raw boned animals.
(4) It must be heavily fleshed, especially deeply down the hind leg.
(5) This flesh must be meat and not fat.
(6) It must produce 'finished' lambs when used with commercial ewes at a wide range of slaughter weights.
(7) Killing-out percentage must be high.

This, then, is the objective. All the recording and the performance figures and the visual judgments must have this picture always clearly in mind. Easily said, far less easily done!

Growth rate is easy to measure, but in a meat animal we must be careful to distinguish between growth which is affected by the milking capacity of the mother and that growth which is clearly the genetic potential of the animal itself when fed on normal forage. Thus we weigh, at birth of course; then at 56 days, which is mostly a measure of the influence of the ewes' milk yield; and finally at 110 days when we get a true and overall picture. There would be something to be said for a further weighing, a year later, at the shearling stage. But this would slow down the generation turnover.

The evaluation of carcase quality internally is much, much more difficult. We slaughter selected animals from each family to cross check on progress. But that is not positive selection. We are now at the stage of beginning to use the Ultrasonic Backfat Scanner. If we can make the same progress with sheep that has already been made with pigs, then we shall really get somewhere.

Finally, we use all our finally selected ram lambs on our own commercial ewes. I believe that this is a vital part of the progress that we make with the Meatlinc selection programme. For we have a thousand ordinary commercial ewes, of the same general type that our customers are likely to have, and these ewes are all crossed with this year's crop of ram lambs. So we have an annual judgment in our own commercial lamb crop, year by year, of what we have been trying to do – and this judgment is backed up by an abattoir assessment.

An Assessment of the Meatlinc

It is perhaps not for me to make any judgment of the success of the Meatlinc to date; or to make any predictions of whether it will make an eventual contribution to British sheep breeding on any real scale. I am far too close to it to do so objectively. I must be a biased commentator. Having admitted that, I may perhaps permit myself the luxury of claiming that at least the logic behind the Meatlinc breeding is sound. A single-minded devotion to market quality without the distraction of needing to keep to certain points of breed type. Yes, it *does* matter what the sheep looks like. And it is important, of course, that the breed is uniform and recognisable and repeatable. These things are not incompatible with the commercial objective. But breed type as it has come to be defined by some of our breed societies seems to me to have little connection with economic reality.

So the market must judge. And happily at the moment is judging favourably. I have always been determined that that is where the judgment, and hence the sales, must come from. The bally-hoo of an extensive promotion and publicity campaign is better left to others.

Chapter 4

BREEDING

THE ART AND THE SCIENCE OF BREEDING

I DO NOT believe that the art of breeding sheep, any more than cattle, can be passed over to what I would call the industrial geneticist. This in no way is to denigrate the science of genetics, nor to suggest that the breeder should not avail himself to the absolute maximum of the help of a computer. It is to say that because of the very low individual productivity of the ewe and the slow generation turnover, we are stuck with a job which is very long term. No industrial company, making annual assessments of its progress as shown up on its balance sheet, can keep its financial director happily interested in a long-term sheep-breeding project. Let us accept that sheep are not pigs and poultry and that the sheep breeder, whatever his difficulties should not look to industrial money for help and support.

I have argued in the last chapter that there is a great deal of breed improvement work required in the two key areas: of the Longwool ram and the Down ram. We are fortunate that we can so concentrate our efforts and it should not be beyond the wit and capacity of the industry to devise some structure which will help to bring improvement about.

What is required of the successful breeder? Patience certainly together with the vision that will sustain him when the going gets rough and the goal suddenly seems an impossibly long way off. That flair that will lead to him making the correct choice. An interest and belief in what he is doing that will amount almost to a passion. All these qualities are what one might term the art of breeding – but essential as they undoubtedly are, they are insufficient in themselves and must be combined with, and complemented by, the science of breeding. An understanding of the mechanics of genetics – the measurement of performance both in the breeding animal and its progeny; the keeping and assessment of such records which must mean access to a computer; and facilities for subsequent testing and evaluation of the saleable ram under commercial conditions in different parts of the country – is of the utmost importance.

THE ECONOMICS OF BREEDING

Even then we haven't finished – for the art and the science must be supported by the economics of breeding. The breeder must stay solvent, and a long-term project demands not only patience but money! We need not concern ourselves with the traditional breeder, let us say of Suffolks or Border Leicesters, for he is breeding and selecting within an accepted and marketable breed. Provided that he is working with a good stock, and does his job only reasonably competently, then at the very least he has a market for his rams for commercial crossing. It certainly will not be an exciting return, but at least he will have an income and, if he is ambitious, he can look forward to making his name at shows and eventually getting into the higher-priced market.

When I say that we need not concern ourselves with the traditional man, I'm really talking about his pocket. What should concern us is whether the progress that we, as an industry, quite obviously need can come out of flocks and breeders who are operating in such restricting circumstances. For once they move away from the ordinary and the traditional, then they are producing a product which will almost certainly not be saleable for a good number of years. In my case, as with the Meatlinc, the breeders' passion must ride on the financial back of other and profitable enterprises such as arable cropping The passion must, in a sense, remain a hobby. The problem is how to maintain the dedication, which this hobby demands for success, without compromising the efficiency and profitability of the arable farming. Alternatively, progressive breeding must depend to some extent on either cooperative or state support.

THE DETERMINATION OF EXCELLENCE

Breeding is all about discovery and the use of superior animals. Fine: but superiority for what purpose, and superiority determined by what methods?

For What Purpose?

More so than with any of the other species, sheep breeding has suffered from an absence of any definition of what the market really requires. This is particularly so with the meat market – the butcher purchaser has never, certainly until the mid-70s, demonstrated his clear preference for one quality of lamb rather than another. All types of lamb could be sold with very little premium or penalty for the best or the worst; and it was quite clear that the most profitable thing any producer had to do was sell his lambs at as heavy a weight as possible with precious little worry about their carcase qualities. Happily, under the influence of the export trade, this is now changing, and there is the beginnings of a premium market for the best quality lamb. It all boils down to money. At the end of the day, it will be the market that determines what the breeder achieves – whether that market be the deadweight purchasing abattoir, the fatstock auction, or the farmer breeder buying a Longwool ram. But that market must be influenced, and it can only be influenced by knowledge, by fact, and by, if I may say so, bringing light into darkness.

Here is where the Meat and Livestock Commission (MLC) has an important role to play. First of all, in the field of different systems of lamb production on the farm, the costing and recording schemes operated by MLC have identified those factors which influence profitability. This work must be developed so that there is a clear feedback to the key breeders influencing their selection. It is difficult to over-emphasise the importance of this flow of information and hard fact. The gap between the best and the worst, be it in commercial production as measured by lambing percentage, or in pedigree ram flocks, is so great that if we could do no more than lift the performance of the average up to that of the top third, we should have achieved a remarkable degree of progress. And, secondly, there is much to be gained by the actual measurement of carcase quality, via the MLC Carcase Classification scheme, so that everyone knows what everyone else is talking about. Then when we talk about fat cover and carcase conformation, it will really mean something which can be translated, *forward*, by the trade into premia and penalties, and *backward* by the breeder into his selection programmes.

By What Methods?

Traditionally, superiority has been determined by the show-ring. The standards used by judges at shows have reflected those currently in vogue and laid down by the pedigree breed societies. In theory, at any rate, these standards are tested year by year against market choice as expressed by the commercial purchaser.

In other words, with beef cattle and sheep, the top herds and flocks have based their selection programme on a visual assessment of animals carefully and expensively prepared for the show-ring. And in the short term, considerable financial reward has come to those breeders who have succeeded. Whilst their animals have been sold to other breeders, the effect quickly travels down the line to the commercial meat producer.

One need but take one extreme example to illustrate the illogicality and the folly of a total dependence on this means of selection. Between the wars we succeeded in ruining two excellent British beef breeds, the Beef Shorthorn and the Aberdeen Angus. Fashion dictated that beef bulls should be square, solid blocks of flesh (fat?)! Standing four square on short legs. Nobody bothered about growth rate, leanness of carcase or even the ability of the bulls to serve cows! But a few, very few, top bulls fetched huge prices at the Perth sales and hit the headlines. The commercial market-place eventually, however, had its say and rebelled against such nonsense. The result was the importation of the French Charolais – very big and lean and an immediate success. There has been a fight back by British Breeders, led by the Lincoln Red Cattle Society who were the first to realise that weight recording of bulls was a valuable tool to use in selection. But, in the meantime, much of the British market has been lost to imported breeds, something which need never have happened.

The contrast between the beef cattle scene and that of the dairy cow is extreme. Selection in performance has led the dairy industry to almost total dependence on the Friesian and more recently the Holstein. And it is no coincidence that of all the livestock enterprises, milk production has consistently remained the most profitable

The examples and the warnings are so clear in the cattle industry that it is really rather surprising that sheep breeders have not followed with greater enthusiasm the somewhat hesitant progress made by their fellow beef breeders. For we have exactly the same nonsense in the sheep world. We have some of the leading Scottish Blackface rams, judged on beauty of horn and colour of face, sold for huge sums; and some celebrated examples failing to breed. Thank heaven that the hard conditions on the hill prevents such idiocy affecting the commercial breed! But then we have Border Leicester breeders enthusing about the roman nose and the length of back. I've never known a roman nose make anyone any money; and I'm convinced that the Border Leicester back has got long to the point of dangerous physical weakness. And so one could go on!

But I must be constructive. Logic leads one to the inescapable conclusion that selection must be based on performance measurement. We are now beginning to have, from the MLC and from other bodies such as the Animal Breeding Research Organisation (ABRO), a clear definition of the specification for the commercial Halfbred ewe and thus those qualities for which we should select in the Longwool ram breeds. There really is no difficulty in laying down a flock selection programme seeking to improve and maintain prolificacy, growth rate, milk yield and longevity. The means of measurement exist, and the results can be built into a selection index. Here let me say that I firmly believe that essential as the use of performance figures are in selection, they must be supplemented by a visual check on physique. Lest anyone should accuse me of being inconsistent, I would emphasise that I am not talking about 'beauty', I am talking about physical fitness for the job.

When we come to the Down breeds, then there are much greater difficulties. The market specification is now much clearer. We can say that we want a ram that, when crossed with the Halfbred ewe, will produce a uniform crop of lambs in the 16–20 kg carcase weight range, heavily fleshed with low fat cover and a high killing-out percentage. What is difficult is to measure these qualities in the live animal. Whereas, with the Longwool breeds, we can measure performance in the ewes and with confidence expect this to be transmitted through their sons; with the meat breeds we have to slaughter before we can measure. We could, of course, use progeny testing and this would be logical in that it is the meat qualities in the subsequent cross-bred lamb that are of commercial interest. But progeny testing is laborious and time consuming and requires organisation on a large scale to put it into effective operation.

What we really need is some means of measuring carcase quality in the young ram at, say eight months old at the time when stud selection is carried out. If the ultrasonic back-fat measurement technique used by pig breeders could be successfully adapted for use with sheep, this would indeed be a breakthrough. It would open the door to selection based on factual appraisal rather than visual judgment of what is rather loosely called conformation, and which in practice is really a description of body fat.

TWO ORGANISATIONAL DIFFICULTIES

Selection apart, there are two organisational difficulties which have to be faced and solved.

The first is the very real physical difficulty of coping with the huge amount of paper-work that a recording and selection programme on any scale produces. Happily, modern technology has produced the computer which can do this for us; and equally fortunately in the MLC we have the central organisation which can carry this out for us. Quite literally, without these central computer facilities a detailed breeding programme would be impossible.

Then, secondly, there is the difficulty of scale. I have already pointed to this fact that in the two main Longwool breeds, the numbers of sheep in each pure-bred flock are very small. There are far too few for it to be possible for these flocks to remain 'closed' – their breeders are bound to go out to buy stud rams if they are to avoid dangerous levels of in-breeding. But to go outside for stud rams negates the whole point of the selection programme, which is to select continuously for the superior individuals. In practice the real influence must come from the male line.

The solution to this difficulty must lie with the adoption, either by breed societies or by groups of breeders, of group-breeding. Briefly, this is an idea which has been developed in New Zealand and subsequently taken up by

ABRO in the UK*. Basically, it consists of a group of breeders pooling their resources and selecting the best individuals for each flock so as to form a separate elite flock. The elite flock is fully recorded and rams for use in members' flocks are selected from it. Quite obviously, such a scheme involves a considerable degree of sacrifice of individuality, and because of this, is not likely to appeal to many breeders whose hallmark, above all else is, their sense of independence. But one has to say that some form of group-breeding scheme is the only alternative to an individual flock size of a minimum of 240 ewes split into four separate families – that is, if one accepts this desirability of having a closed flock.

This is not a specialist book on genetics and sheep breeding, so in these two chapters I have had to pass very superficially over many matters which merit discussion on much more detailed level. On several occasions I have mentioned the services provided by the MLC. I can do no better than refer any reader who is seeking further information on this subject to the MLC Sheep Improvement Service.

*See 'Group Breeding Schemes'. ABRO publication.

Chapter 5

DIFFERENT SYSTEMS OF PRODUCTION

IT WOULD be surprising if in a country as diverse geographically and climatically as Great Britain, there were not to be an extremely variable range of different systems of sheep production. My objective in this chapter is to examine the main alternatives which exist and to discuss their pros and cons. I shall not attempt, however, to lay down a blueprint for each and every set of circumstances. No such clearly-laid-out set of rules can exist: each farm and each farmer has to work out which suits them best.

THE HILLS

I propose to say very little about hill and upland sheep systems, partly because the major aim of this book is to concern itself with intensification of lowland sheep, but more especially because I should rapidly expose the depth of my ignorance. Suffice it to say that I believe that our hill country represents an under-utilised national resource of very great value. One has only got to examine the records of some Scottish parishes of the late 1700s and early 1800s to see how much greater the production could be. Interestingly enough, in those days high production off the hills was achieved, partly by mixed stocking with cattle as well as sheep, and partly by a system of away wintering known as the Shieling system. This quite simply recognised the fact that hill grazings produce an abundance of grass during a short growing season of five months and nothing at all during the other seven. If the stock are to stay on the hill the whole year, then the stocking rate can only be determined by what the hill will carry during the winter. Inevitably, this leads to serious understocking during summer and a consequent and equally inevitable deterioration in the quality of the grazing as the coarser plants thrive and dominate. Thus a vicious circle gets under way, made all the worse by the absence of the cattle who can eat the coarser grasses.

This fundamental fact, that the whole year's output is determined by the winter carrying capacity, is so similar in principle to the lowland situation (which leads logically to inwintering) that a naive lowlander must pose the question – why not inwinter hill sheep? Indeed there are those who do just that, and I remember being very impressed by Mr Jim Lindsay's inwintering of Scottish Blackface ewes at Carmacoup in Lanarkshire.

Of course, I fully realise that there is much more to it than just putting up the necessary buildings. The biggest limitation to any improvement and intensification of hill farming lies in the famine for capital that is triggered off by any one item of improvement. Fence the hill and improve the control of grazing – and you need more sheep. Selectively improve the pasture whether by draining, liming and/or reseeding – and you need more sheep. Mix the grazing – and you need to buy cattle. It seems to me to be so relatively easy on many hills to carry out dramatic improvements and in consequence to run headlong into very serious cash flow difficulties. All this in an environment where the output of the individual ewe must be severely limited even under the improved conditions. One can understand why hill farmers are cautious indeed about any exhortations to expand their output.

Nevertheless, the Hill Farming Research Organisation (HFRO) on its three farms in Scotland, and the Ministry of Agriculture on its experimental husbandry farms at Pwllpeiran in Wales and Redesdale in Northumberland, have carried out work of the greatest value. The way forward for an increase in production off our hills is reasonably clearly mapped out. All that is required is the financial encouragement. Whether that will be forthcoming remains to be seen.

THE LOWLANDS

Perhaps the first distinction to make is to divide those systems where ewes are lambed from those where no lambing is involved.

(1) Lambing Flocks

which should be sub-divided into:

(a) (i) early lambing;
and
(ii) late lambing for either finished lamb or store lamb production.

And either of these may be:

(b) (i) self-contained breeding their own replacements;
or
(ii) purchased anually.

And equally any of these may be:

(c) (i) large flocks managed as a main enterprise either in a grassland or arable context;
or
(ii) small flocks managed in a secondary role as scavengers behind a main enterprise, often dairy cows.

(d) Pure – or cross-breeding for a specific purpose.

(a) Early Lambing v. Late Lambing

The essence of success of early lambing must be to get all the lambs away sold to the butcher whilst the price remains high, certainly by the end of May and preferably by the end of April. Young lamb meets a luxury trade during the early months of the year, when it is in short supply and in competition with the remaining few of last year's wether lambs, and with frozen New Zealand imports. The price is high but it quickly falls away once greater numbers of new season lambs come on the market. A situation which is very comparable to the production of early pototoes.

High cost is inevitable, as the milking ewe has to be hand-fed for a long period and the lamb must be creep-fed, but it is not a system where you can afford to economise. Equally, it is a system where you just cannot afford to miss the high price market. For these reasons it seems to me to be only really suited to those favoured areas where the grass grows early and costs can be reduced; and often on small farms and with small flocks where minute attention to detail is possible. It is also well suited to those farms with mixed stocking where the burden can

9. Stanhope Farm – a typical Scottish Hill farm.

Source: ABRO

10. The Lowlands. Some of the sheep flocks at Worlaby (May, 1979) stocked at 30 ewes and their lambs per hectare.

Source: David Lee.

be taken off the summer grazing, as only the dry ewes will be left on the farm.

Lambing late – to coincide with the onset of grass growth, say end of March or the beginning of April – is the opposite in almost every way. It involves keeping the costs of the in-lamb ewe down to a minimum, with the milking ewe and her lambs being dependent almost entirely on grass. It is lamb production from grass and forage crops and depends for its success on maximum exploitation of these crops.

Costs must be kept low, for there can be no question of hitting the early market. On the contrary, the bulk of the lamb crop must be sold in late summer and autumn, just the period when the biggest weight of British lamb is on the market and the prices therefore at their weakest.

Depending upon the farm and the area, and perhaps most of all on the ability to grow catch crops for autumn forage feeding, these lambs may be kept on into the winter. They will have been on the farm a long time, but if they have been really heavily stocked at grass, they may well profit from feeding on into the winter on roots, by which time market prices will have begun to rise again.

Lastly, there is always the possibility that these lambs can be sold as stores for others to fatten. And this alternative is most likely to appeal in those areas of poor autumn growth and particularly the all-grass upland areas where forage crops can only be grown with difficulty. Also, where the preparation of the ewes for tupping must not be compromised by having that year's lamb crop still on the farm.

All in all, late lambing is likely to be the choice of those outside the early favoured areas; particularly those with large flocks and who are specialist sheep/grassland managers.

One thing about which I am completely certain – those who are neither early nor late lambers hit the worst of both worlds: the high cost of one with the low returns of the other.

(b) Self-contained v. Annual Replacement

The big argument in favour of a policy of purchasing replacements for the flock is that it takes advantage of all the benefits of stratification. One will be using Halfbred ewes with the increased productivity due to hybrid vigour, and there will be the undoubted benefit of sheep coming from a poorer harder land so that they improve in their new home. And finally there is the point that the Halfbred ewe will have been produced with a single aim in view – to produce a good productive ewe; and there will be no confusion created by any attempt to feed in carcase qualities in addition to these maternal ones.

These are weighty arguments in favour of going annually to buy some 20 per cent or so of one's total flock numbers. There are disadvantages, notably two of great importance.

Firstly, there is the matter of cost. The purchaser is exposed to the ups and downs of the autumn ewe lamb and gimmer trade and he may in some years find himself paying out a good deal more than he has budgeted for. Perhaps more significantly, he will not himself be benefiting from an inflationary increase in the value of his flock – he will be paying that over to someone else. This has been a factor of very considerable importance during the mid-70s when we have seen a tremendous escalation in the value of breeding sheep; and in consequence the capital value of a flock has trebled in six years. For those who were already committed to annual purchases, there has been a strong temptation to compromise. The result has been a big jump in the number of so-called Suffolk Halfbred, *i.e.,* (hill ewe x Leicester ram) x Suffolk. In other words, the ewe lamb product of the Suffolk cross onto the Halfbred ewe has been kept on for breeding as a lamb mother and crossed again with a Down ram. I believe this to be false economy. Hybrid vigour in the second Down cross is virtually lost and prolificacy and growth rate suffer as a result.

Secondly, there is the risk of introducing disease. This must always be so, and the risk is particularly high where the sheep are bought from sales and where they are mixed from several different production farms. Scrapie and abortion are perhaps the two most serious hazards and impossible to detect in such circumstances. And the less dangerous but potentially costly diseases such as foot rot, orf and pneumonia are a near certainty. Can these dangers be minimised? The answer must be yes, and to differing degrees.

If you must buy at a sale, then I believe that you should only buy those sheep about whose farms or origin you have some certain knowledge. If you are not able to do this yourself, then it is far better to commission a reputable dealer to do it for you. Better still by far, in my view, is to by-pass the sale ring and buy direct from the farm of origin.

There is a great deal to be said for building up a contact to the benefit of both parties. It may cost the purchaser rather more than if he were bidding skilfully for the lower priced lots at auction – but that is a small price to pay for security. Another advantage can spring from this association – the purchaser has the chance to influence the producer in his choice of Longwool ram, both of breed and quality within breed.

If the decision is to go for a self-contained flock, then the choice of breed is forced upon you. To use but one pure breed on a lowland farm is bound to involve compromise – there is no breed in the UK that meets all the requirements of maternal productivity and produces the best of lamb carcases. In some specialised circumstances, perhaps the Dorset Horn for out-of-season lambing, and maybe the Clun Forest in its lowland form where intensification of stocking is not required, the use of the single pure breed as a commercial lamb enterprise is acceptable, but in general it has too many disadvantages. However, there is another way out of the dilemma which has some appeal, and that is to produce one's own Halfbred ewes out of purchased draft hill ewes. This does not avoid purchasing altogether but it can be limited. And it can be much more secure both from the point of view of health and of costs.

The Scottish Blackface ewe drafted, after four crops, off a sound hill and from a reputable breeder can, with proper management, give at least two crops at surprisingly high percentages. Crossed to a Longwool ram and you have the production of your own Halfbred ewe lambs. It adds complexity to the flock management, but provided the hill ewes are managed with skill and can be fitted into the overall farm planning, then there are considerable attractions. Of course, other hill breeds are used in the same way, notably the Welsh Mountain.

Large v. Small Flocks

This is not primarily an argument about size of farm. It is much more a question as to whether sheep should fill the main and perhaps only livestock role on the farm, or whether they should be supplementary to, say, dairy cows. It is a decision which will be determined in many cases by the character of the farm and its fixed equipment, and very much by the character of the farmer. So perhaps in many ways the element of choice does not exist. Nevertheless, there are many farms where a small flock of, say, 100 to 150 ewes could be fitted in to profit. For example, there are many mixed farms in the Midlands or the West where such a flock could scavenge behind arable crops with very little reduction in overall arable acreage. All that is required is ingenuity in growing catch crops behind, say, winter barley coupled with the skilful use of areas of grassland, which most such farms cannot avoid. Similarly, there are many dairy farms with room for a flock that can utilise corners of the farm that are difficult to get at with the cows. And usually such farms have grass that needs clearing up in autumn after the cows have come in. But skilful management is needed so that the first priority of producing milk from the cows is never affected.

These are examples of where the particular flexibility of the sheep can be used to maximise farm profit. The essentials are not only ingenuity and skill, but also enthusiasm for sheep that is sufficiently disciplined so that the main enterprise is complemented and not threatened. For all these reasons, the small scavenging flock is much better suited to those areas where livestock knowledge is traditional and ingrained. It is definitely not a system for the main arable areas where a small flock would usually be neglected.

It is my experience, as an arable farmer farming a reasonably large acreage, that enterprises on such farms have to be big enough to be specialised, otherwise they just do not function. The small bits and pieces rarely succeed simply because no one is really responsible and no one can justify devoting any time to them. Such arable farmers have to become expert specialists. That should not preclude sheep from their consideration, but it does mean the flock should fill a primary role and be big enough to justify the employment of a full-time shepherd. And that necessarily means a minimum of 700 ewes. This does not infer that the flock should not be intergrated into the rotation, making maximum use of arable by-products – very much the reverse.

Arable farms are, of course, not the only category of farm

which merit consideration of the large specialist flock. There are those mainly grass farms, on which dairying for many reasons may not be suitable, where the same considerations apply. The sheep flock may well be the main enterprise, with beef cattle occupying the supplementary role

(d) Pure or Cross-breeding for a Purpose

The discussion on all these various alternatives so far has assumed that ultimately the lamb is destined for the abattoir. There are, however, many situations where flock profit is better served by concentrating on the production of breeding sheep. The prime example are those farms situated on the more favoured slopes of the uplands, not in the restricted conditions of the hills yet obliged by their environment to use a hill-bred ewe, and yet under conditions which can produce something bigger and better than the hill wether lamb. Under these co ditions the obvious choice is the planned production of half-bred ewe lambs or gimmers.

Then finally, there are those who can choose to become pure pedigree breeders. In every case, this is the choice to be determined by the man rather than by the circumstances of his farm. To be a successful breeder is not really a system, it is a vocation.

(2) Non-lambing Flocks

which we can sub-divide into:

(a) fattening purchased store lambs;

(b) buying ewe lambs to sell as gimmers.

Lambing a ewe flock demands skilled attention and management at many stages in the year and, of course, particularly at lambing time. The success of the whole enterprise depends on how well this is carried out – a statement of the obvious. It is equally obvious that there are many farms where this standard of management is not readily available and yet where sheep obviously have a role. These are the farms that should consider one or other of the following alternatives:

(a) Fattening Purchased Store Lambs

Each autumn there has to be a very heavy trade in store lambs, coming from those farms and areas which for some reason or another cannot fatten them. These lambs may be wether lambs of the pure hill breeds; they may be the wether lamb by-product of the Halfbred ewe lamb producer; or they may be Down crosses of both sexes and out of all breeds and cross-breeds of ewe. Each has a trade, and each has a particular purpose for which it is best suited. Some will already be big and only require finishing and must not be held for too long. Others, such as the hill wethers will be slow to finish and will probably have to be held well into the New Year.

By definition, the fattening of these lambs is short term (on the farm for no more than four months or so) and carried out on those farms that have food available during autumn and winter. By tradition, this was the enterprise for an arable farm growing forage root crops of turnips, kale and swedes as part of the rotation. Times, however, have changed and the justification for the main line root crop as part of the rotation has largely disappeared. But this is not to say that the arable farm cannot and should not fatten large numbers of store lambs on catch crops such as forage turnip and rape or on by-products such as sugar beet tops.

It is an enterprise that has much to commend it on the farm that cannot provide skilled shepherding and yet which has *cheap* food resources available during the latter months of the year. There is, however, one skill which is essential: as with the fattening of store cattle, as much of the profit with sheep is due to skilful buying and selling as it is to management.

(b) Ewe Lambs to Gimmers

Many of those lowland producers who purchase their replacements annually prefer to do so in the form of gimmers. There are a variety of reasons for this, not all of them particularly good ones – but it is a fact. Equally, most of the farms in the main areas where Halfbred production is the main enterprise are obliged to sell their product as ewe lambs. There is a gap to be filled here.

There are those who are, as it were, the professional producers of Halfbred gimmers which have been brought in as ewe lambs. An alternative to the fattening of store lambs, it also avoids the complexities and management demands of lambing. But, unlike store lambs, the sheep are on the farm for the whole year. It has many attractions, and there are areas like the Yorkshire Wolds or the chalklands of Wiltshire where this has developed into a local industry. Significantly these are arable areas and, in the main, large farms.

As with store lambs, buying and selling skill is of the utmost significance; and the production of a high quality and well-presented product is equally important. It is definitely not an option for the amateur – which in a way the store lamb job can be. It demands professionalism of a high degree, but nevertheless is an enterprise which is more easily moulded and fitted into the more industrialised management of the big arable farm. It can use permanent pasture intensively in a way that ewes with lambs never can due to the greater resistance of the adult sheep to worms. It can use grassland of all types very intensively and fits well into the catch crop and by-product potential of the arable farm. It can be attractively profitable, but particularly so in times of inflation where next year's gimmer is as a result worth relatively more than this year's ewe lamb. But any prudent manager should distinguish, in his judgments, between an inflationary profit and a real one.

Chapter 6

MANAGEMENT OF AN INTENSIVE LOWLAND FLOCK

THIS IS the story of my sheep enterprise at Worlaby in North Lincolnshire. I do not suggest that in any way it is out of the ordinary – quite the contrary – but it has shaped my thinking and is therefore the basis for this book. It is also the account of a flock built up and turned into a profit-earning enterprise during a period when it was generally recognised that sheep were not easy money-spinners.

THE EARLY YEARS

The story starts in 1960 when, as I have said earlier, I came to the conclusion that Worlaby required grass in the rotation, and sheep were the chosen means to utilise it. Two-year leys seemed to be about the right proportion of grass, and so over two years we built up to 500 acres plus a small amount of permanent pasture. Those were the days when the Clun Forest was popular, so, following that lead, they were the choice and we increased to a total of 1,200 ewes.

It was all very traditional, following the thinking of the time. A grassland breed of sheep, lambing towards the end of March, producing fat lambs off grass and all stocked at 6 ewes per hectare. I think it is worth recalling that that was about the extent of lowland sheep thinking in the early 60s. Apart from a few researchers looking at the Finnish Landrace breed, there just was not any constructive or imaginative thinking going on. There was just a calm acceptance that this was the only way to do it, and it would not be very profitable anyway.

Well, my word, it certainly was not profitable! At 6 ewes per hectare and a lambing percentage sold of 140, and an average lamb price of £5 in the summer (rising to £7 in the winter), the total gross return was just over £49 per hectare, including wool. Not even in those days did that begin to cover fixed costs, never mind anything else.

It was all very depressing because, no matter how we struggled and schemed in those first three years, I just could not see a way forward into profit. We were stuck with low output, of which there were two components – lambing percentage and stocking rate. It was absolutely fundamental that if we were to stay with sheep, this output had to be pushed up and pushed up dramatically. But how?

Lambing percentage, the productivity of the individual ewe, seemed the obvious target. If only we could move the 140 up to nearer 180 per cent – and some people were talking of those sort of figures. But the demoralising fact of the matter was that we did not seem able to go beyond an honest 140 per cent of lambs sold relative to ewes put to the tup. Were we especially incompetent; or was the flock size of 1,200 ewes unmanageably too big; or what? A nasty suspicion comforted me that in all probablity the high figures claimed were not even true. But, in any case, it did not really affect the issue for at 180 per cent we would have only grossed £67 per hectare – still not enough.

So what about the other component of output: the stocking rate? Six ewes to the hectare – that is to say, six adult sheep as counted on January 1st relative to the total area in fodder crops excluding by-products – was the absolute limit for Worlaby. This limit was imposed by the winter. Even at that figure, there were times in the winter in wet weather on our heavy land when that figure looked dangerously high. The sheep were suffering and, just as critical, the grassland was being badly poached. The consequence was that grass growth in the spring was desperately slow and it was nearly midsummer before it recovered from its winter punishment. But, even so, we had a surplus of grass in the summer and it was completely clear that we had the ability to increase the output of grass many fold.

We were at a standstill and there just did not seem to be any way out. By 1963, I was very close to giving up, and I think I would have done if I had seen any clear alternative.

THE BREAKTHROUGH

Then, however, fortune intervened, and for reasons that had nothing whatsoever to do with sheep, I went on a farmers' study trip to the arable farming areas of northern France. There, for the first time, I saw sheep being housed. I cannot recall what that trip cost me, but there is no doubt in my mind that it was one of the best investments I have ever made.

A fellow member of that group was Eric Carter, then our senior district Advisory Officer of NAAS and since Deputy Director of ADAS. He and I looked at these sheep and the penny suddenly dropped! If our stocking rate was limited by the winter, why not take the sheep off the grass and put them indoors and then we could really go to town with grassland management in the spring and summer. But perhaps French sheep were somehow different; their English cousins might succumb to every disease under the sun. *And* no one else was doing it. (I have long since learnt not to be discouraged by that!)

Eric was, and remained, a great source of encouragement and inspiration. At the same time, I met Dennis Hurst, Principal of the East Riding College of Agriculture, who had studied inwintering of sheep in Norway and Finland. Quite coincidentally, he was thinking along the same lines as I was, and we were a great help to each other in the developments that followed.

Anyway, the upshot of all this was that we carried out a trial with 100 ewes during the winter of 1962/63. This did not appear to present any problems, so I jumped right into the deep end and put up a new building designed to inwinter the whole flock of 1,200 ewes. The winter of 1963/64 saw our whole flock inwintered for the first time. Never for one moment have I regretted it. It was quite literally the key that opened the door and made all other things possible. What this led to is illustrated in Table 6 which shows the progress the flock has made.

THE FUNDAMENTAL LESSON

I shall make no apology in this book for repeating ad nauseam: *the only way to profit with sheep is by pushing up output per hectare.* No matter how one twists and turns, the ewe is basically a low output animal. The only way out is via high stocking densities. The achievement of this in most, but perhaps not all, situations is made far easier by the adoption of inwintering.

Table 6. Progress of Worlaby Flock

	Ha/forage	*Ewes*	*Number of ewe hoggs*	*Rams*	*Stocking rate*	*Gross output/ha*
1961	182	700			3·8	£38·30
1962	203	1173	120		6·4	£39·78
1963	242	1480	189		6·6	£70·42
1964	207	1699	202		9·1	£103·78
1965	190	1450	235		8·8	£106·25
1966	97	1291	430		17·7	£168·02
1967	108	1183	528		15·7	£185·32
1968	118	1394	477		15·9	£190·26
1969	115	1063	586		14·1	£150·23
1970	123	1137	844		16·3	£195·95
1971	118	1105	940		17·0	£222·63
1972	133	1062	1045		15·8	£264·39
1973	158	1193	1592	59	18·2	£513·96
1974	172	1119	1506	106	15·8	£317·63
1975	200	1693	978	97	13·8	£268·92
1976	194	1713	941	158	14·5	£474·05
1977	195	1639	718	132	12·8	£531·61
1978	194	1649	566	159	12·3	£440·35
1979	135	1707	571	176	18·0	£671·85
1980	133	1379	443	149	14·8	£564·00
1981	120	1317	598	111	16·8	£899·00
1982	84	875	419	84	16·4	£908·00
1983	95	1033	374	100	15·8	£770·00
1984	95	1029	433	125	16·7	£893·00
1985	100	1090	410	137	16·3	?

High output via high stocking rate plus a reasonable lambing percentage: adapt that, as you will, for your own farm, but the basic fact stands out clearly.

There is nothing new in all this – the acceptance of this basic fact is merely a return to those principles which our grandfathers knew and practised so well. For what else are we talking about other than the old folded flock, part of the arable farm and contribution to its fertility, *and* very heavily stocked indeed. There was no more intensive system of sheep production than this and all that needed doing was to adapt it to modern costs and conditions. More and more, as I developed the flock at Worlaby, I realised that this fact was central to success. Solving the winter problem enabled us to make a start – after that we had to re-learn many other lessons.

THE GRASS

Suddenly we had to become grassland managers. Slowly edging our way forward in 1964 and 1965, I became aware that the mainly Italian ryegrass type of ley for dairy cows was no good for dense grazing with sheep. Sheep are not at all the same kind of grazers as cattle – they are 'nibblers' and they prefer and do better on fine leaved grasses, close to the ground and forming a dense carpet, and freshly grown. Furthermore, somehow we just had to get this dense carpet quickly: to achieve during the two-year life of the ley what had presumably taken generations on the famous pastures of the Romney Marsh. Then there was the problem of keeping the grass growing in the summer to support what is effectively an increasing stocking rate as lambs grow.

11. Improvement of soil structure. Heavy clay land at Worlaby after only one year in ley. Deep penetration of fibrous roots is already very evident.
Source: David Lee.

None of the grass seeds mixtures on the market at the time met any of these requirements. Neither the simple dairy cow mixtures based on Italian, nor the complex general-purpose mixtures based on nearly everything. We tried them all, and in the end came down to the extreme pasture types of perennial ryegrass. There is no better grass for sheep, of that I am quite convinced. We started with *S23* and have since moved to *Melle,* only because there is some basis for believing that it is more winter-hardy. The only possible disadvantage is slowness to get going, and I have always thought it worth while to include a small quantity of Italian just to give the ley a kick-off in the first spring.

Our mixture since then has been the following:

20 kg	*Melle*
6 kg	previously *S22* Italian ryegrass and now *Sabrina*
26 kg	per hectare

No white clover you will notice, not because I have any feelings against it, quite the contrary. But I have never succeeded in getting any variety of white clover to contribute anything on our heavy land. It establishes itself and then does no more. I think there are several reasons for this. Our heavy land is not particularly suitable for clover. Nitrogen levels at 187 kg per hectare are high enough to discourage it. And most important of the lot, the dense carpet of these perennial ryegrasses just simply strangle the clover. No matter, the carpet is much more important than the free nitrogen.

Only on one occasion was I tempted away from this simple grass mixture. Seed prices had become high and I succumbed to meanness and substituted *S321*, an erect type of ryegrass, for *S23*. The net result on those fields was a fall in stocking capacity of 5 ewes per hectare. False economy indeed, but I learnt a lesson.

THE PLACE FOR ROOTS

So we learnt how to master grassland management for sheep. But we were then confronted with another door closed in front of us. First the winter, then getting the grass right and now the problem of the 'summer gap'. Under the economic pressures of the mid-60s, a good number of people were pushing stocking rates up to try and get those extra few lambs to counteract rising costs. Like me, many had found that it was relatively simple to put on an extra ewe, to go from five to seven ewes per hectare or even rather more, and in consequence produce more lambs. However, it became obvious that it was one thing to produce more lambs, it was another thing altogether to produce more *good* lambs. There was no point in producing more if at the end of it all you had produced a lot of poor quality store lambs.

I well remember the *Farmers Weekly* talking about what they called 'July disease'. Lambs which had done well on their mothers up to the end of June, started to go back in July and become stunted and poor – a condition from which they never really recovered. The critics pointed to this as proof that heavier stocking rates of sheep at grass just were not a sensible proposition.

This never seemed to me to be a very helpful attitude, and my mind went back to the example of the folding flock. Surely these lambs were suffering from malnutrition. The natural curve of grass growth, however much it is flattened and prolonged by good management, dips down in July. Furthermore, ewes which have lambed at the end of March are coming to the end of their lactation by July 1st; and instead of being of benefit to their offspring, suddenly become competitors with them for a declining supply of grass. Throw in an increasing worm burden for good measure, and it is hardly surprising that one runs into trouble in July!

The solution must be to separate mother and offspring systematically on July 1st. There should be no compromise with this rule. But to put the lambs onto what? Good and ample supplies of nutritious, and above all, clean fodder – of course. But again, what? The standard answer was onto hay aftermath. But in a wet season, haymaking wasn't finished so weaning had to be delayed. In a dry season, there was precious little regrowth after the hay, so there wasn't enough food to wean on to. Hardly the basis for precise control.

Roots must surely be the answer – but *roots!* They had

equally surely been discarded in the 30s. Not a very up-to-date solution! And yet, of course, this was nonsense; precision drills and modern herbicides made roots an easy crop for any arable farmer to grow. And, coincidentally, at about that time two new developments came on the scene. The quick-growing forage turnip and the electrified plastic fencing net. Much more about this in a later chapter, but suffice it to say here that was one of the breakthroughs – perhaps the important one alongside inwintering – that took us successfully up to 15 ewes per hectare. So we had secure supplies of fodder, properly programmed to suit the needs of the lambs from July 1st to finishing; not a replacement for grass but an essential complement to it in any intensive system.

THE RIGHT SHEEP

So far in telling this story of the Worlaby flock, I have said nothing about the sheep themselves except that I started with the Clun Forest. I was following fashion. But fashion changed and by the mid-60s, the Scotch Half-bred was popular, to be superseded today by the Greyface, the Mule and the Masham, or even the Suffolk Half-bred.

Fashion, of course, is a fickle guide and the Clun certainly was not the right breed for intensification at Worlaby. It just doesn't like living in large numbers and at close quarters with its fellows, particularly at grass. As we pushed up stocking rates, so we ran into trouble with the Clun. Ewe depreciation went up rather more steeply than did our stocking densities – and this was due not to any particular disease, but was marked by a general inability to 'do' and thrive. Too many ewes consistently lost condition and had to be culled.

This looked like another closed door on the way forward. The clue how to open it came when, very early on, I noticed that some Suffolk × Clun ewes that we had kept on from ewe lambs did not react in the same way at all. They kept their condition, even got fat on the same food, and, most important, were noticeably quieter and much less nervy than the Cluns. Was this due to the infusion of 'Down' blood from the Suffolk, a breed accustomed over the generations to close folding? Back to folding again!

My thinking on these lines was very much stimulated at this point and subsequently by Stephen Williams, the Director of Boots farms, who had given a lot of attention to animal psychology. It was surely illogical to use an upland breed of sheep for lowland intensification. The very qualities which serve a breed so well when foraging in the more difficult and more extensive conditions of the uplands are the very ones that were not wanted at Worlaby. The 'prick ear', the nervous disposition, merely led to something approaching a nervous breakdown when crowded together – and as we know in human medicine, nervous breakdowns can and do lead to physical breakdowns, which was just what was happening to our Cluns.

So we wanted a calm, almost stupid sheep that liked living in a crowd, and would stay happily where you put her, eating without question the food on offer. Above all, she must not be able to jump! There was really only one British breed that met this specification and which had, over many generations, been accustomed to high stocking rates – the Kent or Romney Marsh. A breed which had all the virtues save one, that of prolificacy. How sad that over the years the Romney breeders have concentrated on the export trade for heavily woolled sheep and have in the process run down their breed's prolificacy to what are quite uneconomic levels.

Having said all that, the Romney is a remarkable sheep in that it will thrive happily at extraordinary densities and seems to enjoy it. It seemed right, therefore, to build on the Romney and this is just what we did. First of all by producing a Half-bred via the North Country Cheviot ram on the Romney ewe. A ewe which we called the Romney Half-bred – and whose production was formalised and disciplined by the Romney Half-bred Sheep Breeders' Association in Kent.

The Romney Half-bred filled a useful role whilst we were creating something better; it gave us a lambing percentage of 150 against the pure Romney's 125, and provided it was stocked really heavily, it was a profitable sheep to keep, very hardy and with a low depreciation. But it was always a compromise until we had something with the same qualities but capable of achieving 170 per cent. This we eventually had with the Oldenburg ram producing the so called Oldenbred Half-breds. The Oldenburg is a long wool breed with its origins in northern Germany on the marsh grazings close to the Baltic. It is very similar in many ways to the Romney with the same placid temperament and ability to thrive at high stocking rates on permanent pasture. Where it differs is in its prolificacy, which as a pure breed can be a realistic 200 per cent. These qualities were recognised by two leading breeders in Kent, Hugh Finn and Rene Regandanz, as a means of up grading the Romney. The importations which they organised established the breed in the South-East but sadly their ambitions to improve the Romney were frustrated by the entrenched traditionalism of the Romney Breed Society. I was fortunate to be associated with Finn and Regandanz at that time, and we decided to move into the production of commercial Half-breds out of the hill breeds – in other words using the Oldenburg as a substitute for the Leicesters, producing a Half-bred as prolific as the Greyface and the Mule but with an improved temperament. This was a logical development and it certainly worked. The Oldenbred Half-breds are attractive sheep and fit well into the management of a flock seeking for high output from grass. But they have the disadvantage that their prolificacy is limited to a ceiling of around 170 per cent. And the hard fact of life is that that really is not enough under current economic conditions. So we have to move on to something else.

THE MARKETING PATTERN

I shall have a great deal more to say in a later chapter about future marketing of quality lambs, and how I think we should be getting the best return from them. But here is the place to make the point that at Worlaby we have never tried to produce out-of-season lambs. I firmly believe that we should accept the limitations and profit from the advantages which one finds on one's own farm.

Worlaby is not an 'early' part of the country. There is just no point in trying to produce early lambs anymore than there is in trying to produce early potatoes. We should have all the costs of early production but we would miss the high-priced market. On the other hand, what we have got is a part of the farm which is on an easy working sand and where catch crops of turnips and rape can be grown very easily after early crops of vining peas or winter barley. This gives us an abundance of food in the autumn and that plus sugar-beet tops really determines when we should market our lambs.

That is not to say that no lambs are sold during the summer. Anything that is at the right weight and is ready should be sold, and more and more we are finding that, as we improve the growth rate and easiness of finishing in our Meatlinc sires, we are selling a high proportion in summer and early autumn. This might seem to run contrary to making maximum use of our food supplies in the autumn; but, of course, we can and we do fill up as necessary with purchased store lambs from elsewhere. Indeed I feel sure that one of the major roles of the arable areas should be to fatten hill store lambs during the winter, keeping the market supplied until new season lambs appear. This must become more important, and more attractive financially, as the influence of New Zealand lamb, which is at its peak in the New Year, wanes under the influence of their search for other markets in the Middle and Far East.

THE FUTURE AT WORLABY

Most of the technical problems posed in the developing years can be said to have been mastered. We know that we can stock at 15 to 16 ewes per hectare and produce 22 lambs sold per hectare, and what is more we can be reasonably sure that we can achieve these targets consistently year by year. With the lamb prices which we have had in the late 1970s and early 1980s under the Common Market Sheep Regime, this has been a profitable enterprise. The Cinderella days for British sheep seem but a distant memory – a fact which is reflected in the big expansion in the national flock since the early 1970s.

There is a temptation to sit back and feel satisfied. That would be both complacent and dangerous. For one thing, economic conditions are hardening against us and costs continue to rise at an alarming rate. For another, we have surely reached a peak in lamb prices. Whilst we were passing through the five-year transition period up to full membership of the EEC Sheep regime, we were assured of a 20 per cent increase at least each year. That is now behind us and sheep prices, like those of every other product, will be exposed to the pressures of financial discipline which are now in vogue in the Market.

What line then are we to follow? Surely not that of sitting back accepting that profitability must fall! I doubt very much whether it is really practical to think of pushing stocking rates up much further. That is not unless we considered irrigating the grassland. It is certainly true that there are periods in every year when grass in under drought stress and where irrigation would increase production – but at what cost?

A more encouraging avenue must be with lambing percentage. I have to admit that, over the years, we have had great difficulty in exceeding a real 'sold' percentage of 160. There are many who claim a consistent figure of 175. Why is it that we miss out on this highly profitable extra 15 per cent? Our management and shepherding are neither particularly brilliant nor are they bad. It is true that we are operating with large numbers and, in the process, attention to small detail is difficult, especially in times of stress. I feel sure that we have got something to learn here. How to master the problems of scale and do it with employed labour – but learn it, we certainly must.

There is surely a parallel here with the growing of wheat. We used to be satisfied with an average of 5 tonnes per hectare and were thrilled to bits to get 7! Now we need 7 tonnes to break even and we are all pushing hard to average 10. Already we have fields doing 10 tonnes per hectare, as we have ewes consistently doing 200 per cent, and we keep pushing forward.

As there has been great technical progress in wheat growing, so there has been with sheep management, particularly with ewe nutrition – a subject to which I shall return in detail later on. Suffice it to say here that we can now contemplate, with some considerable confidence, the management of a significant proportion of triplet births which is the inevitable consequence of going for the 200 per cent sold performance. That must be our sheep ambition to match the 10 tonnes wheat yield! To do it, we have certainly got to change breed, either to a pure breed or a half-bred that is capable of 230 per cent live births. The Leicester crosses cannot do that which is why at Worlaby we are, in 1985, moving over to Cambridge half-breds out of the Scotch Blackface. It remains to be seen how we get on. The first indications are that we shall have plenty of lambs. The challenge will be to match our management to the genetic potential of the sheep. There is certainly no reason to be complacent!

APPLYING 'WORLABY' PRINCIPLES ON OTHER FARMS?

On many occasions people have said to me, 'what you do at Worlaby is fine but . . .', and the 'buts' that are cited include: Worlaby is a better farm with rich soil; you have the advantage of operating the sheep within an arable rotation which includes early harvested crops; and, most flatteringly, what you achieve is due to your own particular ability which other lesser mortals have not got.

Let us deal with these, because if they are accepted then there really is not any point in writing this book. And let us dispose of the last one first of all, as that warrants no serious consideration. I am not even particularly close to the management of our sheep, relying on the interest and skill of our shepherds. And whilst they are good at their job, they would be the first to say that they are neither more nor less exceptional than I am. No, if I have contributed anything out of the ordinary, it can only be the fact that I took the plunge in the first place and stuck to it in face of the many discouragements that lay in our path.

So what about the other two. Worlaby is certainly not an exceptional farm, indeed it has several disadvantages, notably the extreme heaviness of most of its land. It is true that it contains something less than one fifth of its area in sand, which in contrast to the clay is easy to work and is particularly suited to the growing of catch crops and for folding sheep in autumn and winter. But there are many, many farms, that are as good or better.

It is true that we have a considerable advantage in respect of rotation in that we live in an area close to freezing and sugar-beet factories. I am bound to admit that that does help in that we have land clear from early peas by the first week in July and can immediately get a crop of turnips in. That gives us a full month's advantage over anyone who has to wait for winter barley before sowing his catch crops. And then there are sugar-beet tops, a really valuable by-product giving a large quantity of food in autumn and early winter. But at the very worst, not having these advantages can only mean that a rather greater area of forage roots have to be grown, with a consequent reduction in the stocking rate of no more than from 16 ewes down to 14 to the hectare. And before we get too excited about this, let us remember that there is a very large slice of the country where peas and sugar-beet are grown and where the same conditions exist as at Worlaby. I would go further than that and say that it is quite astonishing to me that so many people miss out on the possibilities of exploiting the sheep feed that these two crops can lead to.

BASIC PRINCIPLES FOR PROFITABILIITY

I want to reiterate, to re-emphasise that the basic principles on which a profitable lowland sheep enterprise can be built are these:

(1) High output from what is a basically low output animal can only be achieved by very high stocking rates. And that there is no reason why this should not be achieved relatively easily, and with reasonable lambing percentages. There is no need to go for exotic levels of production, although no one can rest content with average levels of performance. The challenge must be to get lamb sales as near to two per every ewe in the flock as possible, and at that level, lamb production can be very profitable indeed.

(2) These high stocking rates are only likely to be achieved on most farms if the sheep are taken off the grass in the winter. In most cases, this means going indoors, but some farms will have other means of achieving the same object.

(3) Grassland management must be first class and the leys must be made up of the right grasses. After all, we are talking of summer stocking rates of around 25 to 30 ewes, plus their lambs, per hectare – that warrants a bit of effort.

(4) Solving the 'summer gap' by growing root crops onto which the bulk of the lamb crop can be put during the critical months of July and August.

(5) Having the right breed for the job. Don't get seduced by the argument that you can stock more heavily with small ewes, simply because their maintenance requirement is less. Irrespective of size, what you must have are stupid, placid sheep that love living in a crowd, and cannot jump fences.

All farms, like all farmers, are different and there can never be a blueprint. Nevertheless, it seems quite clear to me that these basic principles remain valid for all lowland situations; and also indeed, although with some change of emphasis, for many upland farms too. Nor do I see any fundamental difference in interpretation as between grassland west and arable east.

Another more general objection, frequently raised by my fellow arable farmers, is that seeking for levels of high output with sheep demands a high degree of skill in management. Yes, no doubt it does but then so does growing potatoes or sugar-beet, or keeping a herd of dairy cows. Why should anyone think that sheep should be any different in this respect? Yet this is often the case and has perhaps been one of the main reasons why sheep husbandry has lagged behind dairy and pig husbandry; that sheep are scavengers and do not merit serious attention. If this is your point of view, then it should not surprise you if your results are some long way from being satisfactory.

Whatever you think or do, do not shelter behind the excuse that what we have done at Worlaby is in any way special or different. Anything we can or have done, you can surely do just as well if not better.

Chapter 7

INWINTERING

SHEEP so they say, 'are hardy animals equipped with a magnificent coat whose natural habitat is out of doors on natural grazing; and it must be wrong, even unhealthy, to bring them into a building'.

Surprisingly, there are still a lot of people who think like that; even producers who have long since accepted the economic realities of putting their other stock under cover at least during the winter months. Instinctively, many farmers shy away from the thought of inwintering sheep and do so without even looking at the arguments. If you are numbered amongst these, my aim in this chapter is to convince you that you are wrong. For if you are serious about sheep, about making money from sheep, particularly on high cost land, then it is my view that you just cannot afford *not* to inwinter.

So what are the arguments in favour? First and foremost, it is certainly not to protect the sheep; on the contrary it is all about protecting *the grass from the sheep.* And that becomes more and more important as stocking rates are pushed up. Sheep damage grass during the winter in two ways. First – obviously – by poaching the land and breaking up the sward. Secondly, and just as important, by the constant nibbling of the young grass shoots, particularly in February and March; they exhaust the growing power of the plant and seriously retard real growth in the spring.

The major benefit, then, from inwintering is that the door is opened to exploit to its maximum the potential of the grass during the growing season; *and* (tremendously important) to be sure of ample grass during the critical month of April. It is difficult, if not impossible, to over-emphasise the economic importance of this. To be seriously short of grass in April is to compromise the ewes' lactation unless generous hand feeding is carried out; and once a ewe has lost her milk, she never fully regains it. It is hardly necessary to underline the fact that lambs' growth and vigour depends entirely on this milk supply. A poor milk supply at this critical time and you have stunted lambs – nothing is more certain.

Of course, these grassland benefits can be obtained by 'off' wintering rather than 'in' wintering. The essential is to separate the sheep from the grass. There are those farms that have some otherwise unusable piece of land where the sheep can be wintered – for example, a steep but sheltered valley. There are other circumstances where it may make sense to make available an area of 'sacrifice' pasture which will be ploughed up in the spring to go into some crop. But if this means growing spring barley instead of a good crop of winter wheat with a consequent and considerable loss of income, then it is very doubtful if it makes any economic sense.

But let me repeat, and underline, the fundamental reason for inwintering: *to maximise the production from your grass.* You will only do that if you increase your stocking rate per acre. Inwintering is no magic formula for making money from sheep. It is a means of producing more grass and thus keeping more sheep, selling more lambs per acre and as a result, making more profit.

POINTS IN FAVOUR

There are, of course, other reasons in favour and let us look briefly at them:

(1) Protection from the weather. The long winter of 1978/79 made many people think seriously of bringing their sheep indoors. All right, it may only be a once-in-a-while occurrence, but when a bad winter comes it can be both a powerful and an expensive argument.

(2) Real control of nutrition. You will know exactly what your ewes are getting, and furthermore can adjust the ration with precision according to stage of pregnancy and condition of the ewe. You can do this out of doors, of course, but indoors you ewes will be in a more stable climate. They will not be exposed to the extremes of driving rain, cold winds, warm days – all of which place differing demands on the ewes metabolism, and the

end-result is a fluctuating level of *net* nutrition. I have always believed that this is the basic reason why twin lamb disease (pregnancy toxaemia) is virtually unknown amongst inwintered ewes.

(3) Production is definitely higher. When the ewes were tupped outside, how can inwintering affect individual ewe productivity, you may ask, apart from being able to save more lambs at birth? We now know that the ewe behaves like the sow in one respect. If she is under stress during the first two months of pregnancy, she can absorb a foetus thus reducing a multiple birth to a single, or in the worst of cases to nothing.

(4) Avoiding waste of food. With properly designed and constucted racks, wastage indoors should be next to nothing. No matter how good the equipment outside, waste is inevitable under bad weather conditions. And then there is the damage done to wet land by taking tractors out each day to feed.

(5) The ability to sort the ewes into different categories. Gimmers and ewe lambs separate from older ewes to avoid bullying. A pen of 'old girls' who have lost their teeth and who would normally be culled can be kept on for an extra and often very productive year if on their own and given some favoured treatment. And, of course, separation, lot by lot, according to date of lambing.

(6) Last and by no means least – the ease and comfort of shepherding. Maybe it is difficult to put a figure on it, but a good shepherd doesn't come cheap and he is bound to be able to do a better job for you under reasonable conditions. The shed ought to be near his house to give you what I call the 'carpet slipper' advantage – he can pop out last thing at night to see that all is well. And all this is a hundred times more important during lambing.

It all adds up to a formidable list of arguments in favour – so formidable that it is perhaps surprising that so few people have taken the plunge. One of the reasons for this has been that the economics just do not make sense unless the stocking rate is consequently pushed up, and there is more, much more, to higher stocking rates than just simply inwintering, as we shall see later in this book. Much of the advice, therefore, coming from both ADAS and MLC has been extremely cautious – much too much so, to my mind, but one can understand the reason why.

POINTS AGAINST INWINTERING

It is only fair to say that there are – not surprisingly – some points against inwintering. Primarily, one of capital cost. This can be infinitely variable. A sheep building does not have to be complex, indeed it must be simple, and there are many farm buildings which can be adapted quite cheaply. Nevertheless, coming indoors costs money; and at the very worst, with a brand new building on a green field site may cost up to £40 per ewe housed at 1984 prices. So this decision has to be weighed very carefully – more of this later.

Then there is the matter of disease. Undeniably if you bring your ewes into a building for the winter months and always lamb them there, you have a disease risk which is greater than if you can choose a different lambing fold each year. A greater risk, yes, but an entirely manageable one and I can honestly say after 22 years of inwintering that it has never caused us any serious worry.

Finally, there is the question of food and bedding costs. It is said that sheep cost more to feed indoors than outside because they are entirely dependent on hand feeding. Quite right, because outside they are nibbling off the young grass shoots which are the 'factories' waiting to produce your April grass. One of my major reasons for coming indoors.

I have devoted a lot of space to a consideration of the pros and cons (quite rightly so, I think) so that you may clear your mind before passing on to some of the detail.

THE BUILDING

(1) *The Principles*

I have said already that the building should be simple; but there are some basic principles which, under no circumstances, must you sacrifice.

First and foremost, the need for ample ventilation. It used to

12. Inwintering shed at Worlaby empty during the summer. Showing design of hay-cum-concentrate racks. Note ample ventilation and Netlon side netting.
Source: David Lee.

13. New inwintering shed built by Mr. Alun Evans, Cwmbur, Mochdre, Newtown.
Source: Farmers Weekly.

14. Cattle fattening shed successfully adapted for inwintering ewes at Druids Lodge, Wiltshire.

Source: Author.

15. Meatlinc cross lambs being fattened on forage turnips at Worlaby.

Source: Author.

be said that bringing sheep into a building was asking for pneumonia. I can honestly say that we have never had a case of pneumonia which could in any way be attributed to housing. But the key is the ventilation. If the atmosphere is stuffy, or there is condensation, or worst of all if you can smell ammonia, then your ventilation is woefully inadequate and you are in for trouble.

Fortunately, we are not talking about expensive fans, and insulation and keeping the animals warm. We are housing sheep not pigs! Remember we are bringing the sheep indoors to protect the grass not the sheep; and all we really need is a roof. Walls? Well certainly not floor-to-eave walling, but it is better to avoid floor draughts, so some form of walling is advisable up to say, 1¼ metres high. Only if the site is really exposed is it necessary to provide anything higher than this. The problem is really one of rain and/or snow blowing in and wetting both the sheep and, more seriously, the bedding. A simple and relatively inexpensive cure it to use what I call plastic netting – Netlon – a strong perforated plastic sheet which can be attached to a supporting structure to form the enclosure. It has the great advantage of allowing both air and light to pass through, but stops virtually all rain and snow. Yorkshire boarding is a very good solution, but a very expensive, albeit longer lasting, one.

To sum up on ventilation; too much rather than too little. The only danger of pneumonia should be to the shepherd and not to the sheep.

Having got that right, then your building design must pay attention to two other matters – the sheep management and economy of labour.

For the sheep, it is highly desirable that you should be able to subdivide the flock into manageable units. I believe that this should not be much more than 50 or 60. The point is that you should be able to sort your flock into its different age, health and condition categories; and then having done that, you should be able to 'see' individual sheep. This business of being able to see individuals is very imortant and is surprisingly difficult in large numbers, particularly when they are packed in close together. Then there is the question of bullying – large numbers together lead to dominance developing with the inevitable result that some will suffer.

There are those who say that, taking all these reasons into account, 15 to 20 ewes per pen is the optimum. I cannot agree that it is necessary to go to such lengths when inevitable cost per ewe housed must shoot up. Indeed we have successfully housed ewes in a shed for 15 years now where, because of the dimensions of the building, we have to put 150 ewes in each lot. But we have had to select with great care.

And the *labour economy*. We really must pay great attention to this, even at the cost of a more expensive building. I know that I regret bitterly that my two buildings are not designed for mechanised feeding. Not only because we are under pressure of escalating costs of which wages are a major element, but also because of the necessity of freeing as much of the shepherd's time as possible so that he can give attention to the detail that successful management of an intensive flock demands.

(2) *Some Essential Dimensions*

A lot of attention is often paid to the space requirements of an individual ewe. According to breed size, figures of between 0·95 and 1·10 m^2 per ewe are mentioned. This is maybe a good guide to the size of building that you will require; what is much more important is the *length of feed rack* available to each ewe. Unlike cows, sheep are flock rather than individual feeders. They must all be able to feed at one and the same time, and if this is not possible, then the weaker ewes get pushed out.

Incidentally, another side effect of insufficient rack space is that those ewes who are pushed out, try to get to the feed by mounting those who are feeding. They scrabble with their front feet on the backs of the feeding ewes – the result is torn and broken wool and this is often the cause of wool loss which is then subsequently blamed onto the 'unhealthiness' of housing.

So rack length is very important, and I suggest that this needs to be 455 mm of rack face per ewe housed. This will appear to be over-generous when the ewes first come in, but as they get heavy in lamb, will then only be adequate. Now this fact has a dominating effect on the arrangement of the pens within the building. In order to combine maximum use of floor space, say, one square metre per ewe, with 455 mm of rack face per ewe, you must have pens which are relatively long and narrow rather than square. This presents some difficulty in designing the ideal building, not least of all because long narrow buildings are fundamentally more expensive than the standard wide span construction.

Of course, a wide span building can be used perfectly well if the pen divisions within the building give you the right combination of dimensions – long narrow pens within a square building. That sounds fine but remember that you will have a lot of feed passages. This will push up the cost by increasing very substantially the overall area required per ewe. Covered feed passages are expensive, and to have several in a square building will increase the captial cost quite unreasonably. The temptation then is to have pen divisions *without* feed passages – double-sided racks forming the divisions and this works well provided hay is the basic food and it is in the form of small bales. This indeed is the case with my first, and biggest, inwintering shed. I can assure you that it is a fundamental error, not for sheep reasons, but that is precludes mechanised feeding. With employed labour becoming more and more expensive and difficult, mechanised feeding is highly desirable.

There is another temptation and that is to say that, for the bulk food, be it hay or silage, it is not necessary to have this full 455 mm of rack. No more than 150 mm is necessary as the sheep will be fed ad lib. Only for the concentrate is it necessary to have the length necessary for all to feed at once – and this can be provided for by putting troughs on the floor. In my experience this is false reasoning. Firstly, because ad lib feeding of bulk foods is quite wrong during the latter stages of pregnancy. Secondly, because concentrate troughs on the floor are obstructions and during the excitement of the daily feed, ewes are jumping over and around them. I am certain that this leads to abortions.

(3) *The Floor*

Don't waste money on concrete, except for the feeding passages. Sheep muck is dry and there is no advantage whatsoever in having anything other than dirt floor underneath it. But perhaps the real argument is whether the sheep should be on straw bedding or on slats.

Properly designed slats (Fig. 2) work perfectly satisfactorily. It is their cost which will put them out of court under all but the most extreme of conditions. And that really boils down to availability of straw. Only in the most remote of hill areas where straw is prohibitively expensive, and timber often relatively cheap, will slats be an economic consideration. As a rule of thumb, slats can easily double the cost of housing, and that expenditure must be very difficult to justify. In my view, slats are not an acceptable alternative. In the lowlands, even if the straw has to be bought, then there can be no question that slats are out of the question.

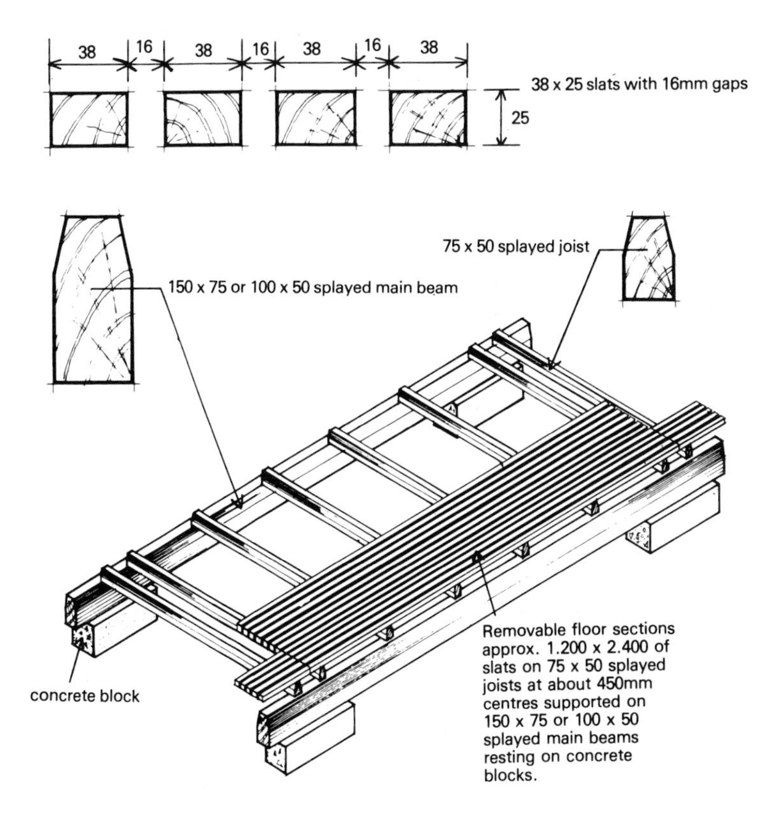

Fig. 2. Suitable type of slatted floor for sheep.

Source: "Housing Sheep", FBC Report 13.

I have mentioned concrete for the feed passage. It all depends, of course, on the design and shape of your building, but where there are tractors and food wagons passing regularly, then concrete must be a near essential. Also, and particularly if it is under cover, the passage is the obvious place to carry out routine treatments such as foot-rot treatment, vaccinating, etc.

(4) *The Equipment*

The feed racks are the most important and can be expensive, particularly if you go out into the market and buy some of the extravagantly-priced and badly-designed ones that are on offer. You will have to live and work with these racks for a long time, so be sure and start off right.

Remember one fundamental principle. The sheep's natural eating position is with nose to the ground – *not* stuck up in the air. Indeed a sheep cannot swallow if it is forced by the design of rack to put its nose up in order to take the food. In consequence, it will take a mouthful, them retreat a step or two in order to bring the food down to floor level so that it can eat in comfort. This can be one of the major causes of feed waste, for much of what is brought to floor level is subsequently trodden on and spoilt.

Sheep are grazers whether they are on grass or fed hay or silage; and the design of rack should take this into account. It follows that the traditional rack shaped, in cross section like an X, is quite wrong. The correct design will depend on whether hay or silage is your chosen food. Silage is easy, basically a trough serves the purpose quite adequately. For hay, the design has to be somewhat more complex. Fig. 3 shows a successful example.

Other equipment is basically concerned with water and light. Sheep drink a great deal of water – it can be up to 2 gallons a day when on a dry diet and is absolutely essential. It must obviously be clean water and it must be kept clean. Sheep are fastidious feeders and drinkers and will always cut their consumption rather than take something they dislike. The standard cattle-type, rectangular water trough, serves perfectly well, and one trough can be situated between two pens.

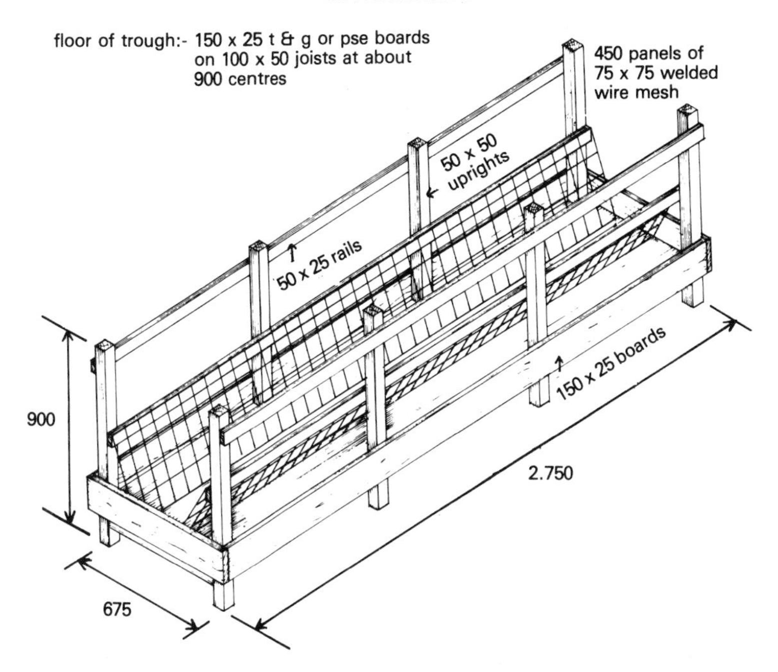

Fig. 3. Good example of hayrack.

Source: "Housing Sheep", FBC Report 13.

Two matters of detail are essential. First remember frost. The building is airy and well ventilated and is therefore cold. It is a confounded nuisance and an awful waste of time to have to thaw out water pipes every cold morning. To my mind the best solution is to have a system whereby you can drain everything last thing at night, installing troughs of a reasonable size so that there is sufficient water available until the morning. It is very rare for pipes to freeze during the day, for apart from the fact that temperatures will be higher, there will be much more water being consumed and thus it will be kept moving.

Secondly, with water, it is very much my experience that it is well worth while to have an area of slats around the troughs. The area need not be big; say, the length of the trough and 1·8 metres back from it; big enough to make the ewes stand on it whilst they are drinking. This is important because sheep tend to splash and dribble when they drink, and without slats you soon have a wet soggy area around the trough. This becomes a focal point for developing and spreading foot troubles.

Lighting: whatever you do, do not skimp it. Ample lighting so that you can really see your sheep is a priceless advantage, particularly at lambing time. It is, I think, worthwhile having the lighting wired to different switches so that you need not have full lighting on all the time. Half illumination which can be left on all night at lambing, and which can be increased to full when required, is an advantage. It avoids the stress of going from pitch dark to full light – stock, as with human beings, just don't like that. Finally, install plenty of power points in the area where you will be putting the lambing pens, principally for infra-red lamps.

SUMMARY

Never forget that you are taking the sheep off the grass. Never fall into the trap of thinking that you are housing an animal like the pig. The essential is cover, adequate natural ventilation, proper feeding arrangements, and convenience of management and labour economy. These principles hold good whether your building is a farm-erected pole barn, and adaption to existing buildings, or is a custom-built permanent building. And a final parting shot – build it so that you can *change* it. The correct design for today's machinery may be all wrong in five years' time.

Chapter 8

FEEDING AND MANAGEMENT INDOORS

AN INCREASING number of farmers are now bringing their flocks inside during the winter. This means that more ewes are becoming totally dependent on the skill and knowledge of the feeder. During the last decade considerable advances have been made in defining the nutritional requirements of the ewe and feeding strategies can now be based on a sound knowledge of the nutritional principles involved in the digestion and utilisation of food by the ewe in late pregnancy and early lactation. Knowledge of these principles is one thing – application is quite a different matter. In the first instance we are dealing with a pregnant animal whose load may vary from one to three lambs and to satisfy such a wide range of nutritional needs is a daunting task. Admittedly the introduction of real-time ultrasonic scanners for the identification of foetal numbers opens up the way for easier and more economical feeding. Any thoughts however that this innovation can substitute entirely for the keen eye of the shepherd should be banished immediately. Inevitably errors in the identification of foetal numbers will occur and the occasional ewe diagnosed as carrying a single will have twins. It will therefore still be up to flockmasters to spot, at an early stage, ewes that are falling back in condition and in need of preferential feeding.

So we are now in a position to base feeding practice on recent advances in our understanding of the principles of food utilisation and of the needs of the ewe. Add to this the experienced eye of the feeder and we should be in a position to ensure optimal feeding regimes, whatever the type of flock we are dealing with.

Most ewes will be coming inside in late November or early

December. In some cases where there is ample food available or where lambing is late, they may be left out until after Christmas. Whatever the actual date may be they will be coming in off grass or perhaps some other low-dry-matter forage or root crop. If it is grass its nutritive value should not be overestimated. As the grazing season progresses the soluble carbohydrate content of grass declines and so too does the efficiency with which the ewe can utilise the energy from grass. The visual appearance of a lush autumn sward can flatter to deceive in a highly dangerous way.

So we ought to think and act on the winter management of ewes before winter sets in and certainly well in advance of the ewes coming indoors. Abrupt changes in the type and quantity of feed in early pregnancy are not conducive to maximum embryo survival. Most sheep farmers are now fully conscious of the need to have ewes in good but not over-fat condition at tupping. Fewer are aware of the importance to embryo survival and therefore to the size of the subsequent lamb crop of maintaining the body condition of the ewe in the first month of pregnancy. As far as nutrition in this vital first few months of pregnancy is concerned the key phrase is 'maintaining body condition'. Extremes of nutrition are detrimental to embryo survival and are responsible for low lambing percentages in both the overfed fat ewe and the underfed thin ewe. To prevent excessive loss of body weight in November and December when the quality and quantity of grass is declining a small amount of hay is ideal. Certainly concentrates are not necessary and hay of moderate quality has the advantage that individual ewes can eat as much as they wish without the risk of digestive upsets at this critical stage in the establishment of pregnancy. This supplementary feeding also prepares the ewes for the change of diet that coming indoors involves.

BRINGING THE EWES IN

If at all possible, choose a day when the ewes are dry. Really wet heavy fleeces take a long time to dry out and in the process make the whole atmosphere and bedding in the shed unnecessarily wet and far too conducive to an outbreak of pneumonia.

I shall be talking about health and disease in another chapter, but this is the point to emphasise the importance of bringing ewes in with sound feet. It is well worthwhile using the bringing-in operation to carry out thorough trimming and tidying up of the feet. I would much prefer to bring in each day only the number of sheep that can be thoroughly well treated.

If you have followed my advice the ewes will already have been started on some dry food. My view is that their ration should be built up to the correct level over a period of a week thus giving the micro-organisms in the rumen time to adapt.

WHAT THE EWE NEEDS

Let us think for a moment about the animal which we are setting out to feed. Firstly, she is small, perhaps between 55 kg to 75 kg liveweight according to breed. Secondly, we are asking her to produce a load of lambs – let us say twins with a total birth weight of quite possibly 9 kg or more – which relative to her liveweight is one and a half times that we would expect from other species. Compare this to the human mother of the same liveweight as the ewe who produces a baby of 3 kg! For such a performance, the ewe merits careful attention.

So we start the winter – say in December – with a small but efficient ruminant who will be perhaps six weeks pregnant. At this stage her foetuses are minute, weighing only about 6 grammes each. On the other hand her placenta or afterbirth should be growing rapidly. This is the organ through which the foetuses receive their nutrients and a well-grown placenta is essential for the birth of strong healthy lambs. Since the entire growth of the placenta takes place during the first three months of pregnancy, ewes coming indoors in the middle of this period must be eased onto a nutritious diet which will maintain them in good working condition during mid pregnancy. By 'good working condition' I mean avoiding both excessive losses and excessive gains in weight. Ewes that are in good condition at mating can loose up to 5 per cent of their liveweight during the second and third months of pregnancy without any detrimental effects. Indeed recent research shows that this may enhance the growth of the placenta but one must be cautious – higher losses of weight have the opposite effect and will impair the growth of the placenta beyond the point where it can subsequently sustain maximal foetal growth. At the opposite extreme, ewes which get too fat in mid pregnancy have insufficient scope left for 'steaming up' prior to lambing and are particularly prone to loss of appetite and pregnancy toxaemia in late pregnancy.

When the ewe arrives at six weeks off lambing, foetal growth begins to accelerate fast and indeed most of the development takes place during the last few weeks. There is rapid transformation in an animal whose digestive capacity is never very great to a state where the ability of the digestive tract to supply the nutrients needed by the fast-growing foetuses is pushed to its limit. And, of course, this coincides with the period when a fall in appetite can occur.

I like to think of this 'steaming up' period in the same way as we do for the dairy cow. The aims are the same. To produce lambs which are strong and viable but not grossly big. And, above all, to develop the tissues of the udder so that the foundation is laid for a high milk yield and a long lactation. The ewe responds in exactly the same way as the cow, and the rewards are as great.

Then finally, lambing and the subsequent turn out to grass. Relieved of her internal burden, the capacity of her gut increases and she can cope with greater quantities of food, but her nutritional requirements remain high – particularly of protein. Lactation involves the highest level of protein synthesis that any animal ever achieves and the milking ewe is no exception. So where must the protein come from? Certainly not from her body. Energy, yes (she will steal from herself by using up her fat reserves) but protein, no. The essence of the matter is this: if there is insufficient protein in the lactating ewe's diet her milk yield will fall to that level determined by what protein there is, and it will do so within three days. Furthermore the quality as well as the quantity is important. And here a word or two about what protein quality to the ewe in late pregnancy and early lactation is all about. Firstly, it is about proteins that are not readily broken down to ammonia by the micro-organisms in the rumen. Currently these are referred to as proteins with a low content of rumen degradable protein (RDP). Secondly, the so-called biological value of the fraction of the protein that escapes breakdown in the rumen (referred to as the UDP fraction) must be high. Fishmeal is undoubtedly the best on both counts, with soya bean a good second. Linseed, which may be one for the future, is also good but rape and groundnut are too high in RDP to be of value.

And so – out to grass. Quite against all logic, sheep seem to accept the transition from dry feeding inside to grass outside without any digestive problems whatsoever. Provided you have ample grass, the ewe and her lambs can go out abruptly and hand feeding need not continue. A word of caution however – when fishmeal is used as the protein source its gradual withdrawal from the diet when grass becomes available is advisable as abrupt removal could lead to hypocalcaemia. In any case the change from dry feed and a benign indoor environment to lush grass and cold wet nights is far too conducive to hypomagnesaemia in some flocks to risk an abrupt withdrawal of concentrates without providing suitable free-access minerals. Grass is a highly nutritious food on which the ewe can build her lactation and feed her lambs. But the grass must be ample, and with the exception of the occasional late spring, we have always had that quantity of grass at turn out immediately after lambing. That is the reward of inwintering.

FEEDING STRATEGY IN MID PREGNANCY

I made the point earlier that ewes in the correct body condition at mating can lose a little condition in mid pregnancy without any detrimental effects on foetal growth. This is the time therefore when savings in food costs can be made. For this class of ewe a roughage with a metabolisable energy (ME) content of 8 megajoules (MJ) per kg dry matter (DM) is an adequate feed during this period. It is important however that the roughage contains enough degradable protein (RDP) to allow maximum growth of the micro-organisms in the rumen and for hays and straws this is not always the case. Quite often they fall below the minimum crude protein content of about 10 g per MJ of ME that is needed. If this happens their intake and digestibility will both be improved by adding a non-protein nitrogen (NPN) source such as urea. But only if sulphur is also added as those micro-organisms in the rumen need, in addition to the nitrogen supplied by the urea, sulphur to ensure that they grow and multiply at their maximum rate. And how much sulphur is needed? Not surprisingly the amount is linked to the nitrogen. Urea solutions containing the correct proportion of sulphur (1 part sulphur to 14 parts nitrogen) are now available for spraying on dry roughages.

In view of my earlier remarks regarding the importance of easing the ewe on to a nutritious diet when she comes indoors around six weeks after tupping I am reticent to talk about feeding lower-quality roughages in mid pregnancy. What I would point out however is that if the roughage falls short of meeting the energy needs of the ewe at this time a small supplement (50 g per ewe per day) of fishmeal in addition to the urea will further improve intake and will enhance not only the protein but also the energy status of the ewe. If these still fail to maintain the ewe there are only two options left. Change to a better quality roughage or introduce a carbohydrate supplement in the form of cereals, sugar-beet pulp or a mixture of the two – not forgetting of course that a little protein concentrate may also be needed.

ENERGY IN LATE PREGNANCY

So much for mid pregnancy. Now its on to the period of rapid foetal growth. This starts 90 days after tupping when the weight of the unborn lambs is still only 15 per cent of their weight at birth some 8 weeks later. It is at this point that foetal growth takes off and imposes a progressive limitation on the use of roughage as the sole feed. This is particularly true for ewes carrying twins and triplets. For such ewes what sort of increase in food needs are we talking about? Well the daily energy requirements, expressed in terms of ME, of a twin bearing ewe of 70 kg at mating increase from 12 MJ at 90 days to 15.2 at 116 days and 18.5 in the couple of weeks before lambing. Using good-quality hays and silages containing 10 and 10.5 MJ of ME per kg DM and a cereal-based concentrate containing 12.8 MJ these target intakes for ME can be achieved by introducing the concentrate at a daily rate of 150 g per ewe as a supplement to 4 kg of silage or 1.25 kg of hay at 90 days and steadily increasing the concentrate intake to 400 g daily at one month before lambing and 700 g just before lambing.

Where roughage is of poorer quality than we would like it is tempting to compensate by feeding greater quantities of concentrates. In theory this appears to make good sense – the heavily pregnant ewe has a high demand for glucose and concentrates tend to enhance the production of propionic acid in the rumen. Propionic acid in turn is the forerunner of glucose. Sadly the putting of this sort of theory into practice can be disastrous. So what goes wrong and why? Well, firstly, we must remember we are dealing with groups of ewes and within the group there is enormous variation between individuals in their rates of eating. This can easily lead to some individuals overeating concentrate, particularly when its usage is high. Need I say more – those of us who have seen the devastating effects of an over-feed of a barley-based concentrate in a heavily pregnant ewe will not want to see it repeated! But even if we do manage to push concentrate intakes up to high levels without any obvious signs of acidosis we may still be making life difficult for the ewe. High levels of concentrates lower the pH in the rumen and this in turn depresses the growth of those micro-organisms that digest the fibre component of the diet. Inevitably this leads to a fall in roughage intake. When this happens the original and very sensible step that was taken of introducing concentrates as a supplement to roughage turns into the absurd situation in which the concentrates substitute for roughage. Forcing the intake of roughage down by feeding too much concentrates or by deliberately restricting its intake to allow high levels of concentrate to be fed sounds sensible in theory but in practice it forces the heavily pregnant ewe on to a knife edge. So much so that it can put her right off feed, which in turn lays her wide open to metabolic upsets such as hypocalcaemia and pregnancy toxaemia – and all because we have failed to take into consideration the optimum environmental conditions for the survival and multiplication of her rumen micro-organisms. If I have conveyed the message that the first step in the successful feeding of the heavily pregnant ewe is the maintenance of a healthy environment within the rumen then I have suceeded. In doing so I hope I have made it clear that the provision of high-quality roughage and a modest amount of concentrates is a better feeding strategy than poor roughage coupled with excessive concentrate usage.

PROTEIN IN LATE PREGNANCY

So much for energy in late pregnancy but what about protein? The extent to which the concentrate requires to be supplemented with protein in late pregnancy largely depends on the level of energy intake. For ewes meeting their energy requirements from the diet, the protein produced by the rumen micro-organisms (about 8 g per MJ of ME) is adequate to meet the protein needs for foetal growth up to the beginning of the last month of pregnancy. Thereafter diets require supplementation with conventional protein sources, in particular those with a high UDP fraction.

The need for supplementary protein in the diet increases as the gap between ME intake and the energy needs of the ewe widens. Bearing in mind that in many situations it is difficult to meet the energy requirements of high-producing ewes in late pregnancy, particularly if the roughage is not of the highest quality, it is desirable to introduce a small amount of supplementary protein about six weeks before lambing and gradually increase its rate of inclusion as lambing approaches. High protein 'balancers' for grain, based on fish-meal or soya bean meal or a mixture of the two with added minerals and vitamins are ideal. A guide to the inclusion rate of for example a 'balancer' based on fish is about 5 per cent of the cereal component of the rotation at six weeks before lambing rising to about 10 per cent in the final couple of weeks for ewes with an expected lambing percentage of 200. Due to its higher degradability in the rumen a higher inclusion rate is required if soya bean is to be the protein source. The great advantage of a balancer is that it adds flexibility at the individual farm level in that its inclusion rate can be varied to meet the specific needs of the flock.

If we have got the feeding strategy right in the run up to lambing the result is strong lambs that are quick to their feet. But more than that, thanks to that supplement of high-quality protein in her diet the ewe will have an udder bursting with colostrum.

LACTATION

Hopefully within a day or two following the lambing it will be out to grass. But what if ample grass is not available? Certainly we should not underestimate where the milking ewe stands in the production league. With two lambs at foot, each growing at a rate of about 300 g per day which is what we expect, she is as productive in relation to her size as the dairy cow producing 30 kg of milk daily. To achieve this level of production without loss of body tissue, daily intakes of 30 MJ of ME (equivalent to three times maintenance) are required. I admit that in practice these are seldom achieved. This is particularly so if ample grass is not available and ewes have to be held indoors. So what can we do? Here we can call on the fat reserves of the ewe. If we have got our feeding in mid and late pregnancy correct then three-quarters of the 20 kg of fat that was present in the body of the 70 kg ewe at mating should still be there at lambing. At least half of this fat, equivalent in milk production terms to the energy supplied by 50 kg of the ewe's diet, can be used for milk production for the first 6 weeks of lactation if the need arises. It can only be used efficiently however if the diet is supplemented with a high-quality protein.

HAY

So much for the needs of the ewe and the principles of nutrition that are involved in meeting them. But what about the specific ingredients? For bulk foods there is a choice between hay and silage, or some combination of both, and the extent to which straw can form some part of the ration. Both hay and silage can be entirely satisfactory, but with the essential proviso – that the quality must be good. Perhaps the first thing to do is to sort out in our minds the pros and cons of each.

Sheep really love good hay; and if the choice were to be left to them, then there is no doubt that it is hay they would go for. But what good is hay in this context? Above all it must be of high digestibility, low in fibre, cut young and made well. A demanding specification indeed! Let us admit straight away that consistently good hay is much more difficult to make than is good silage. But 'difficult' is not 'impossible', particularly if you are prepared to treat haymaking as a job which merits highly professional and skilled attention. There are those people who make good hay with impressive regularity, but they lavish on it the sort of care that many top potato growers apply to that crop.

This is not a book to go into the techniques of hay, or silage, making. Suffice it to say that two relatively recent developments have made quality haymaking a lot easier. First, the use of additives which, in my experience, definitely do control the development of mould. And secondly, the use of barn hay drying, so that the period of weather risk in the field is cut to the minimum. One method that combines preservation with increases in both digestibility and intake is of course ammonia treatment. For hay the increases in digestibility and intake are not nearly as large as with straw but nevertheless provided the treatment is cost-effective and can be fitted in on the farm it is worth considering.

The problem with hay, to my mind, lies with handling it rather than making it. Never mind what the enthusiastic advocates of big bales, be they round of square, may say – the fact remains that hay handling cannot be mechanised in the way that silage can without involving high capital cost. The only way that you can avoid putting a great deal of human physical effort into the handling of hay in the summer is by using big bales. Somewhere along the line, particularly if you wish to barn dry, you have to do some hard work. I am sufficiently Victorian in my outlook to believe that that never did anyone any harm, but the fact remains that the provision of hard work is both expensive and difficult. Furthermore, big bales do not fit in well with the concept of barn drying which, in my view, is almost essential. On the other hand, one has to say that for winter feeding, and particularly for different buildings, the small hay bale has a great deal to commend it.

SILAGE

What about silage? Do not let us fall into the trap of thinking that making good-quality silage is child's play either. It certainly demands the same dedication that the top-flight milk producer puts into it. For there can be even less room for compromise with quality than there is with hay. Silage for the pregnant ewe must be the product of a good fermentation, it must be highly digestible and of a high dry matter. It is just no good producing low dry matter silage for sheep. Remember the tendency for the ewe's appetite to decline as she advances in pregnancy. We just cannot afford to fill up that limited digestive capacity with either water or indigestable fibres.

I have seen far too many cases of ewes actually near to starvation on rumens full of totally unsuitable material. Perhaps even more dangerous, because it is much less easy to see in the ewe carrying a heavy fleece, is the insidious loss of weight that marginally bad food can lead to. The result is pregnancy toxaemia, lambs born weak and unviable, and ewes lambing down with little or no colostrum. This situation makes a mockery of vaccinating the ewe against the Clostridial diseases so that protection can be passed on to the lambs in the colostrum!

Happily it is quite unnecessary to make poor-quality silage. Given good-quality grass to start with, cut at the correct stage of growth, almost certainly wilted, and then treated with an additive, silage-making has become a precise technique.

So, what about the choice? Quality silage is perhaps rather easier to make consistently than quality hay. But silage makes heavy demands on capital, though hay is not cheap. Sheep prefer hay to silage and at comparable qualities will consume 15 to 20 per cent more dry matter from hay than silage. So much for the pros and cons. What I believe decides the issue is the question of ease of handling. If you can be equipped with the whole range of silage machinery – from the high-speed mower and the forage harvester to the well-made clamp, to the mechanical unloading of the clamp into a self-unloading forage box, and with buildings suitable for the use of this forage box – if you have got and can justify, by the scale of your operations all this and can take advantage of the dramatic increases in intake that accompany precision chopping then you have virtually 100 per cent mechanised sheep feeding.

With all these qualifications, I would go for silage. Sadly because my buildings are quite unsuited to mechanical feeding, I am stuck with a dry forage and hard work.

OTHER BULK FOODS

Finally, on the subject of bulk foods, what about straw? In theory the choice is between barley, oat or wheat straw but in practice it virtually boils down to barley – oat being a rare commodity nowadays, wheat being low in nutritive value and barley being so plentiful that one can select for quality. And make no mistake about it, ensuring that you select one at the upper end of the digestibilty range is extremely important if you intend to adopt a feeding strategy which relies on straw as the main roughage. Even so, as a feed for mid pregnancy it will probably require supplementation with nitrogen. This will involve adding urea and/or a conventional protein along the lines I mentioned earlier. Even a small amount of grain may be needed to ensure that the ewe's energy needs in mid pregnancy are met. Later on, as the nutrient requirements of the ewe progessively increase, additional quantities of high-quality protein such as fish or soya, plus of course extra energy in the form of grain or sugar-beet pulp, will be required to compensate for the relatively low nutritive value of straw. Careful however with that extra grain – remember what I said earlier about the detrimental effects of very high concentrate feeding on the rumen micro-organisms that digest fibre.

There are of course ways of avoiding the need for excessively high levels of concentrate feeding when using straw. The obvious one is to allow the ewes to select what straw they find paticularly appetising and leave the rest. What remains each day can be forked out to act as bedding. In this way the ewes will have the benefit of taking in a feed which is considerably higher in nutritive value than would be the case if they were forced to clear the remaining poor quality roughage from their troughs. And there is the recent introduction of ammonia treatment, which if properly carried out, enhances the rate of digestion of the straw in the rumen, thereby increasing its digestibility and more importantly its intake. Ammonia treatment has the added advantage of correcting, at least in part if not completely, any deficit of degradable protein.

CONCENTRATES

The choice is fairly straightforward. Do you buy ready-mixed, proprietary concentrates from a reputable firm? Or do you mix your own, using cereals and purchased proteins?

The first is easy and expensive; it is also reasonably foolproof, but makes no allowance for differing quality in the basic roughage element in the diet.

The second is, of course, much more trouble but significantly cheaper. There is, reasonably enough a substantial margin in the formulation, mixing, pelleting and packaging of proprietary foods which you can pick up for yourself. Especially so, if you are a cereal grower as well. The real advantage comes in being able to adjust the concentrate ration according to the changing needs of the ewe and to the analysis of your bulk foods.

Of the cereals, oats used to be regarded as the traditional food for sheep but all that has now changed. Barley, with its higher content of available energy and lower fibre is now the standard cereal. But barley can be a dangerous feed, particularly if hammer milled. Barley meal should be avoided at all costs. If you have to process it at all, then rolling is better than milling in that it interferes less with roughage digestion in the rumen.

Sheep digest whole grain barley very efficiently, particularly when fed in conjunction with hay or roots. Some grains escape digestion when silage is the roughage but even then it is doubtful if the costs of processing are justified. When fed in the whole form the grain interferes less with fibre digestion than when rolled. There is, however, one significant disadvantage in feeding whole grain barley. Any spillage, out of the trough, can lead to barley being spread over the farm where it can become a weed. I have this problem myself; admittedly, much worse when we are feeding lambs outside in the autumn on catch crops. Barley falling into the ground then grows in the following wheat crops. For this reason, I (and I think, most arable farmers) are forced into the expense of rolling.

There is another solution with cereals which I think has much to recommend it. That is wet storage with addition of propionic acid. At feeding you are dealing with a dust-free product which rolls well and is much more appetising. But remember the moist storage of grain causes a deterioration in its vitamin E status and this must be corrected prior to feeding.

The other carbohydrate food available which has a great deal in its favour is sugar-beet pulp. This is, of course, more attractive to the arable farmer who is also a beet grower and who has therefore the right to buy at favoured prices. But even to the non-beet grower, pulp can be an attractive proposition. Basically, it will be a matter of comparing prices to see which is the better buy. But as a food, either in the loose pulp form or as nuts, beet pulp is well liked by sheep and with certain qualifications they do well on it. It is dry, has about four times the fibre content of barley and there does seem to be a definite limit to the amount that a ewe will eat. We have found, with 70 to 80 kg ewes that in the pre-lambing period we must not exceed 0.5 kg/head/day of beet pulp nuts if we are to avoid a reduction in intake of other foods

PROTEIN FOODS

Then what about protein? I have already mentioned the importance of protein quality in relation to feeding in late pregnancy and early lactation. Fishmeal first, soya a good second; but groundnut and rapeseed because of their very high degradability in the rumen, right out of consideration as far as feeding to the high-producing ewe is concerned. Mention high degradability and of course urea immediately springs to mind. It is 100 per cent degradable and can only benefit the ewe when her basal diet is too low in degradable protein to meet the nitrogen requirements of the rumen micro-organisms. I have already discussed the spraying of urea/sulphur solutions on low-quality dry roughages in mid pregnancy. These solutions are now available commercially, as are urea/molasses licks. The same principle is applied in the manufacture of feed blocks except that the basic 'carrier' is a cereal. The commercial range of liquid feeds and feed blocks has recently been extended to include those with some high-quality conventional protein added. This is in recognition of the progressive increase in the protein needs of the ewe in late pregnancy.

Provided the correct application rate is adhered to then the spraying of urea/sulphur solutions on roughage is a safe on-farm operation. This is not the case when urea is being incorporated into a concentrate mix – urea can be a dangerous food to play about with and its successful and safe utilisation depends upon the formulation of the feed. Under no circumstances should any individual be doing his own mixing. Buy from a reputable company that has a successful track record, and my experience is that you will have absolutely no trouble whatsoever.

Are urea foods worth considering? The answer is undoubtedly yes – in the situation where there is a need for a degradable protein. So too are liquid feeds and feed blocks, their main attribute being convenience in certain circumstances. Blocks do not have a place in indoor feeding – on the other hand they can be highly effective in a hill situation. Liquids also have some advantages. They avoid the need to mix protein with the cereals, so there is one job less to do. Feeding is easy via ball-lick feeders. They form an appetising food and ewes take to them readily, no doubt attracted by the molasses.

Finally, a brief mention of minerals. Brief, not because minerals are not important, nor because I have mentioned earlier the need to guard against hypocalcaemia and hypomagnesaemia, but because no one should determine the mineral supplementation for their sheep as a result of reading a book. There is far too much unnecessary mineral feeding. Equally, there are many sheep which suffer, chronically or acutely, from certain mineral deficiencies. There is only one piece of advice to give. Take impartial advice following both soil and fodder analysis.

RATIONING INDOORS

Bearing in mind everything that I have written so far in this chapter, let me try and put it all together in a sort of feeder's guide to the rationing of inwintered ewes.

The ewes are coming in off grass and have been appropriately prepared for this change by the introduction of small quantities of dry food. Once indoors, they should be taken up, over the course of a week, to something approaching *ad lib* feeding. This applies whether they are basically on hay or silage.

Should it be *ad lib*? If the roughage is of good quality then *ad lib* feeding will not be necessary. For poor quality roughage I have already pointed out the advantages of allowing the ewe to select. To do this effectively what remains in the trough at the subsequent feed should be removed before fresh roughage is added. Remember the principle in these early yet critical stages of pregnancy that ewes should not be allowed to get overfat – neither should they be allowed to lose too much condition. Within any flock, at any time, there will be a wide range of condition in the ewes. It should have largely been evened out in the period prior to tupping but there will be some variations nonetheless. So within the fairly narrow band of change in body condition that is desirable at this stage it is essential that ewes are sorted into different condition categories and fed accordingly. This will be in or around the time for scanning for foetal numbers and grouping according to subsequent needs on the basis of litter size can also be made. You just cannot and must not treat a large flock as a single unit for which there is one blueprint ration. So for goodness sake, look at your ewes, feel your ewes, and, if in doubt, weigh regularly a marked selection of ewes so as to see just what is happening. And watch out for individuals slipping back too far in condition because of the occasional underestimation of foetal number. A heavy fleece is a great cover up of losing condition and serious loss can occur before you are aware of it.

If hay is your basic food in this period, then you will have to decide whether it needs any supplementation. If it is late cut, over-mature hay or barley straw, you will have to determine whether there is a case for spraying on a urea/sulphur solution; the early introduction of a liquid feed; or if it is so low in energy, that it warrants a low level of cereal feeding. But if the hay is of the superior quality that you have aimed for, then no supplementation should be necessary. But, and it is a big but, do analyse and know for certain.

This approach then should take us through to the beginning of the run up to lambing, say, six to eight weeks before. Remember that we are now entering the period of rapid increase in the size of the foetal burden. Our problem is to get sufficient highly nutritious food into the ewe and that means progressively increasing the amount of concentrates. We should know by this stage the approximate amount of concentrates that we intend to build up to by the end of pregnancy. If it is 0.4 kg/head/day (say for singles) or more (in the case of multiples) then we should introduce twice-a-day feeding as soon as possible. Also we should make every effort to keep feeding times standard from day to day. Ewes quickly get into the routine of expecting their concentrate feed at a particular time and to be given it at that time minimises digestive upsets.

So the rule of the thumb is this: depending on condition, start six weeks before lambing at about 0.2 kg/head/day of a balanced concentrate. If your hay or silage is high in protein and digestability, you can start with a mixture of 90 per cent barley and 10 per cent soya bean meal, or its equivalent (about half these amounts if it is fish). Your aim should be to increase this to 0.5 kg/head/day by three weeks before lambing, at which point you should increase the soya to 15 per cent and reduce the barley to 85 per cent. During the last two weeks concentrate intake will probably have risen to 0.75 kg/head/day or slightly higher, if roughage quality is poor. But remember what I said earlier about the dangers of very high concentrates usage. For barley-based concentrates I consider 1 kg/head/day, given in two equal feeds and well separated in time, to be the absolute upper limit just before lambing in ewes carrying multiples. Make no mistake about it, this is a massive amount of concentrate for a ewe at that stage of pregnancy.

I can sum up this approach in no better way than describing it as top-flight dairy cow management. And what is the reward? Strong viable lambs and ewes with three-quarters of a litre of colostrum ready and waiting. That may seem a lot but it is what is needed to satisfy the needs of her lambs immediately they are born and also to ensure that they acquire the maximum degree of immunity against disease. Hopefully you will have ample grass to turn out to. You will, however, want to keep the ewe and her lambs penned individually for a day or two; or even in bulking-up pens prior to turning out. Do not stint on feeding immediately after lambing – remember the rapid increase in appetite at the beginning of lactation; remember also the point about the protein. And watch the water – see there is ample clean water at all times.

Finally, a thought about feeding which surely applies to all stock and which I believe to be of the greatest importance: the stimulation and maintenance of appetite. When I discussed silage or hay or straw, I treated them as alternatives. The really good feeder will be seeking to vary the diet of his stock so as to maintain their interest and their appetite.

FEEDING OUTWINTERED EWES

This chapter has been all about feeding the inwintered ewe. I finish as I began by pointing out that by putting stock indoors, you are making them entirely dependent on your skill as a feeder. The principles of feeding them to ensure optimum performance are brought into very sharp focus. But these principles apply equally to ewes outwintered – and indeed perhaps with even greater force as they have to submit to the full rigours of the climate and this can increase in the maintenance needs by 50 per cent.

Much of the technical information in this very important chapter comes from Dr John Robinson of the Rowett Research Institute at Aberdeen. I am greatly indebted to him, not just because of this book, but because he has taught me and many others how the ewe should be fed for maximum production. It is not too much to say that his work has changed my attitude to the preparation of the ewe for milk production to the extent that we can quite happily contemplate leaving triplets on a healthy ewe. And that opens up the prospect of using more prolific sheep. The industry is fortunate to have men of the calibre and enthusiasm of John Robinson working for it.

Chapter 9

GRASSLAND MANAGEMENT

WE ARE not growing grass for fun; or even for the satisfaction of winning a prize in the ley competition organised by the local Grassland Society. We are growing grass because in our climate we consider it to be the best and cheapest means of summer feeding the largest number of sheep per hectare consistent with satisfactory individual performance. And we are working with an animal that is a competent and efficient grazer.

WHICH GRASSES?

Grass, however, is a blanket term covering a multitude of sins. So what is it that the densely stocked flock of ewes and lambs looks for from its pasture? Watch both sheep and cattle grazing and you will be immediately struck by the difference between the two. Whereas cattle wrap their tongues around a bunch of grass and literally pull it off, sheep nibble delicately. This tells us something of fundamental importance – cattle prefer and therefore do best on relatively long grass; sheep loathe long grass and indeed find it extremely difficult to eat.

Secondly, sheep prefer young shoots and leaves and will always select them in preference to older grass. And, finally, not only do they much prefer a dense pasture because they have to cover so much less ground in the course of their grass-harvesting operation, but also it is this type of pasture which will support the largest numbers of sheep.

If you doubt any of this reasoning, then go onto a green moorland – I am thinking of many parts of the Yorkshire Moors that I know well – and notice where the sheep are grazing. You will find them cropping areas that look like beautifully kept, fine-leaved, dense lawns.

If we have any sense at all, we have to respect the sheep's preferences and marry them together with the highest possible output per hectare. These two are certainly not irreconcilable. But I am going to be completely dogmatic and say that this reasoning leads us to one grass and one only – the extreme pasture types of perennial ryegrass; varieties such as *S23* and *Melle* which quickly grow into a dense carpet, never growing to any great height but producing a thick mass of small fine leaves. They give the appearance of not being particularly high yielding, especially when placed alongside an impressive field of rampant Italian ryegrass. I repeat, we are not concerned with appearances; we are concerned with maximising sheep grazing days per hectare, maximising sales of lamb meat per hectare, and that is what these varieties can give you.

What about Italian ryegrass? There is a great temptation to include some in the ley mixture for the sake of earliness and bulk. I think there is some justification for this, especially in short-term leys to fill the gap before the pasture perennials really get established. But only a small amount, so as to avoid the risk of actually retarding the perennials. And for goodness sake, punish it hard and never let it get boss of you.

The meadow fescues, the timothys, the cocksfoots? Forget them is my advice, they just don't come into the same league. They are either unpalatable or uncontrollable or difficult to grow. Why bother with such problems when we have such grasses as *S23* and *Melle*.

Finally, white clover. I am an advocate of white clover, if you can make it grow, but it's not possible on my land at Worlaby. However, there are plenty of soils where it will make a significant contribution both to lamb growth and to saving in nitrogen fertiliser cost. Surely we must put more effort in the future in both plant breeding and grassland management so as to try and get more from white clover. It has been a much neglected plant over the past few decades when the very low cost of nitrogen has put clover in the background of people's minds. Red clover? Don't touch it! Although lamb growth rates look impressive, there is the real danger of hormone levels, which when fed to ewes can reduce their prolificacy very considerably.

Permanent and Temporary Pastures

I have been writing so far as if short-term leys were all we were concerned with. I must put away my arable habits and broaden my outlook! In fact there are only two good reasons why anyone should be interested in, say, two-year leys. The land is quickly back into an arable cash crop, having contributed reasonable benefit to soil structure. And with such short life leys, sheep worm problems are at a minimum. The disadvantage, and it is a considerable one, is that such leys are really very expensive to establish. Seed, particularly of the extreme pasture types which by definition produce a poor seed yield, is costly, and then there is the work of cultivation and drilling however it may be done.

So long-term leys and permanent pasture are not only an intrinsic part of the country's grassland, but they also have very significant cost advantage. Furthermore, with good management, there is no reason whatsoever why such pastures should not be as productive as short leys. Always providing that the correct balance can be maintained in the plant population, and particularly providing that such invaders as meadow grass can be kept out; with these provisos, then permanent pasture can be as productive, if not more so, than the short-term ley. Everything that I have said, however, about the virtues of the pasture ryegrasses still holds good. The aim should remain the same. Whether you have chosen and bought the seed, or whether you have inherited a permanent pasture, your management should be aimed at producing a sward where these pasture ryegrasses dominate.

The Curve of Grass Growth

No matter which species or variety of grass we are concerned with, they all show, to a greater or lesser degree, the same pattern of growth throughout the growing season.

After the winter's dormancy, there is the slow start into growth in the spring; followed by the explosion of growth that peaks in early summer; then it falls away into a trough in late July and August, to be followed by a minor peak (the autumn flush) in September and early October. In some ways, this is very inconvenient, particularly so because the curve runs contrary to the total liveweight stocking rate with sheep, which

16. Mechanised feeding of grass silage to inwintered ewes on a farm in France.

Source: Author.

17. Electrified fence netting used to fold fattening lambs on forage turnips.

Source: Author.

18. The ideal sheep pasture. A dense carpet of short leafy grass.
Source: David Lee.

19. Melle pasture perennial ryegrass being cut for hay at Worlaby, June, 1979. The grass is just into seed head, the ideal stage for making high quality hay.
Source: David Lee.

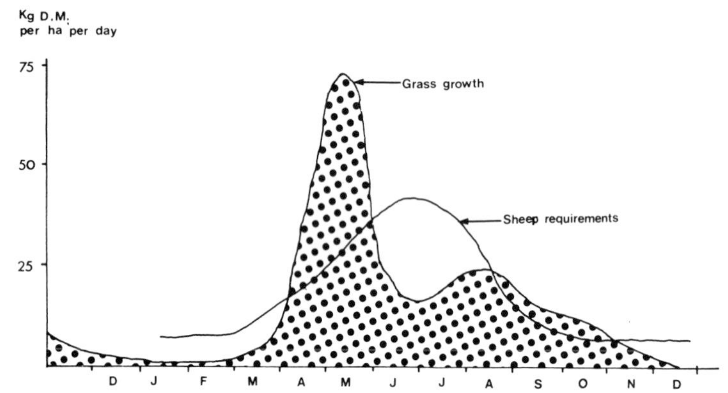

Fig. 4. The growth of S24 ryegrass and the requirements of ewes and lambs (12 ewes/ha: twin lambs).

increases steadily throughout the season as lamb weights increase until such time as significant numbers of lambs begin to be sold.

Obviously, we are in a position where there is not a sufficient supply of food from grass for the numbers of sheep we wish to feed, unless we determine our sheep stocking rate by those numbers that can be carried during the low production period of the summer trough. This would be completely uneconomic, so we have to do something about it. Nor need we consider a stocking rate which is sufficient to consume, *in situ*, the peak of the early summer flush, here is the source of the bulk of the ewes' winter food to be conserved in the form of hay or silage.

We can, however, influence the shape of the growth curve. Firstly, our chosen grasses do help enormously in that they have a very low profile curve – the peaks are not exaggerated, nor are the troughs. In great contrast to Italian ryegrass, where the opposite is true. Secondly, by applying nitrogen at the right time, we can force our grass into extra growth at times when normally this would be slowing down. Thirdly, and most important, the pasture should never be allowed to 'run away'. Grass, like most plants and man, has the instinct to reproduce itself so as to ensure the survival of the species. Once this urge has been satisfied, then growth slows down. Successful grassland management must deny to the component grasses in the sward this basic opportunity.

It must be said that irrigation could technically play an

important role. Although the summer trough is part of the plant's natural cycle of growth, this is exaggerated by high summer temperatures and low soil moisture. Irrigation can help to overcome this, but it is very doubtful if it can be economic for sheep-utilised grass, certainly at present levels of cost and return. It is just perhaps a marker for future development that the New Zealanders are in some circumstances using what they call border dyke irrigation very successfully for sheep. And no one can say that they get overpaid for their lamb!

Then, finally, on the growth curve of grass, we have to admit that, having used all our skills, we are still left with a problem in July and August. The so-called 'summer gap'. There is no way in which the gap can be closed by the use of grass unless uneconomic levels of stocking rate are to be accepted. The answer lies in the use of forage crops, which is the subject of the next chapter.

Table 7. Comparison of Yields of Some Forage Crops with the Production of Grass from Mid-June to the End of the Grazing Season

	Yield per hectare (tonnes)	*Dry matter/hectare (tonnes)*
Grass	14	2·7
Continental/stubble turnips	86	7·7
Forage rape	32	4·5
Kale	44	6·2

FERTILISERS

This is not the book to go into great detail about soil analysis and precise fertiliser applications. Rather is it my aim to set out principles. And so with fertiliser – surely one to get out of the way at the outset is the principle that reserves of phosphate, potash and lime must be adequate. There is just no point in talking about intensive management of either sheep or grass unless they are. Soil analysis followed by the appropriate doses of these essential elements must be the order of the day.

On point of detail, well worth stating because it is a mistake which is made far too frequently: *don't put potash on in the spring*. The evidence is that if you do, you will increase the risk of grass staggers (hypomagnesaemia) in the ewes. Some farms and some soils are much more prone to this than others, but my advice is: don't take the risk wherever you are. And there is absolutely no need to, for both phosphate and potash can be put on very conveniently in the autumn without any risk of significant loss by leaching.

Then, nitrogen. There was a time when there was a school of thought which said that any real quantities of N on grass were dangerous for sheep and in particular, made them more susceptible to such diseases as enteroxaemia. This approach has become much more muted recently as it has become evident that sheep can not only tolerate, but also thrive on grass which has been highly stimulated by nitrogen fertiliser. Having said that, I do not believe that there is any economic justification for going to the lengths used by some of the advanced milk producers who talk gaily of a minimum of 500 kg of N per hectare per year. It is my experience, on reasonably fertile lowland at Worlaby, that 180 kg of N in three separate dressing of 60 kg is quite adequate. These are normally applied in March, May and July. But hard and fast rules are not the essence of the game. Flexibility and adjusting to the circumstances of the year are all part of the skills required. In particular, an anticipation of a likely shortage of grass is an essential weapon in your armoury. It is far too late to be putting N on when you have already run short of grass. How you develop this anticipation, this 'nose' for trouble, is not something that I can tell you. It is only something that you can develop for yourself.

Finally, a word on how and when to apply nitrogen. I am an advocate of set stocking and I have never found it in any way to be a disadvantage to apply the fertiliser whilst the sheep are still in the field. Any idea that nitrogen is nasty, poisonous stuff that will harm the sheep is nonsense. It makes sense to apply it when there is moisture about, for the simple reason that it will be absorbed more quickly by the plant.

SET STOCKING v. CONTROLLED GRAZING

This is an argument where there tend to be passionate advocates of one course or another. There is no such black and white answer so, as with most livestock debates, it is better to go back to basic principles.

Let us recall what it is that the sheep likes. Short, fine-leaved grasses. Old shepherds have a saying that sheep prefer today that grass which has grown during the preceding night. I think there is a lot of truth in that. Watch sheep grazing; watch them on a short, tightly grazed pasture and then watch them being turned into a fresh field where the grass is longer. They will dash around, confused, and searching into the bottom of the sward for the new sweet growths which they can nibble.

So that is one principle of importance. The second, of equal ranking, concerns stress. I say of equal ranking, but it is much, much more than that, for the avoidance of stress is a fundamental of good stockmanship, not just at grass but throughout the animal's whole life. But we are considering grass and grazing systems at this point. Throughout the spring and early summer we are concerned with a family relationship, of the ewe and her lambs; but also a family relationship which has to succeed in a crowd. For if we are to achieve an overall stocking rate throughout the year, including conservation, of 15 ewes per hectare, then at this period we shall have at least 25 ewes and twins per hectare. A stressful situation in itself, and one which has a bearing on breed choice. Add on to that, a separation of mother from children and you have real stress, and you have to see some very significant compensating benefit to justify it.

It seems to me that creep grazing, by its very nature, goes against these two principles. Let us briefly consider how the classic form of creep grazing, as originally developed by Professor Cooper at Newcastle, works. The field is divided into a minimum of six equal-sized paddocks. Grazing of these paddocks passes round in rotation, the ewes staying in each long enough to defoliate the grass completely, normally three to four days. The cycle is thus calculated to be completed every 21 days or so. The basic concepts of this system are these:

(1) The lambs are encouraged to pass into the next paddock in the sequence through creep hurdles, so that they can have the first go at young nutritious grass. These grass leaves are completely unsoiled by the ewes, having grown afresh from a completely defoliated situation.

(2) The lambs are helped to go forward by offering them a palatable creep feed in the first few weeks, but this is

quickly withdrawn. In effect, it is population pressure, *i.e.,* sheer overcrowding, which drives them forward.

(3) The ewes are used as vacuum cleaners, as it were, cleaning up everything that is left behind. And it really must be a clean sweep.

(4) In this way, the lambs get preferential access to nutritious worm-free grass, which theoretically is at exactly the right stage of growth. Furthermore, the grass plant, having been given a rest, has had the time to recuperate and regain its energy and thus produces a greater yield of nutrients per acre.

Certainly, creep grazing is a highly intensive form of grassland utilisation. Large numbers of ewes and lambs can be kept on a very small area. And it can be made to work, though only with some very skilled management. But it has to be said that there have been far more failures than successes, and whereas it was promoted with great enthusiasm in the late 50s and 60s, it has now basically been discarded.

What I did not like about creep grazing was its air of what I call organised chaos. The noise of ewes and lambs shouting and calling for each other was to me a danger signal. A field of ewes and lambs should be like the nursery for a human mother and child, a place of calm. What happens, let me ask you, if a nursing mother is subjected to stress, if you have a row with her? After the next feed, the baby will have the tummy ache! And life is miserable for everbody!

So I tried, and discarded creep grazing. But I discarded it for sheep reasons, rather than from the point of view of doing everything to maximise grass yield. When all is said and done, it is sheep that we sell, not grass. The fact remains, however, that constant nibbling, particularly when it is taken to the point of overgrazing where the plant is taken off right down virtually to the roots, does reduce yield. Grass does require a period of recuperation after times of intense effort. A natural and entirely logical reaction. This is one of the reasons why the extreme pasture ryegrasses are better than any other grass – once established it is virtually impossible to eat them off down to the roots.

Still the nagging thought remains. Is there some way of combining the benefits of set stocking with those of creeping?

The system known as 'Follow N' developed at the National Agricultural Centre at Stoneleigh goes some way to doing this. The essence of it is simple – put N on to grass and for a while it becomes less palatable than grass which is further away in time from its top drèssing. Therefore, in a field which is set stocked, it is possible to top-dress only a quarter, say, of the field each week. There is, therefore, across the field a succession of stripes which are at different stages of nitrogen response and hence palatability. Selective grazing by the flock does seem to follow the top-dressing at the appropriate interval without the need to fence or separate, and therefore with no stress. This is an interesting and intelligent development which merits serious attention.

THE MANAGEMENT OF SET STOCKING

Making the assumption that lambing will be timed to coincide with the onset of grass growth – in our part of the world, at the end of March – then ewes and lambs will be turned out to reasonably adequate grazing. However, British springs being what they are, grass will be in short supply certainly for most of April. It will not be until temperatures, especially at night, really rise that we get that explosion of growth which is such a dramatic turning point. Until that happens, then the flock will need to be dispersed at as low a stocking rate as the total area of grass allows. But as soon as that surge of growth starts, then they must be tightened up, progressively freeing those fields which are destined for hay or silage.

Tightened up to what level? Of course, it is impossible to give a precise answer, only experience can tell. It is, however, very difficult to change the stocking rate once you have got the sheep in the field and each field has its quota. Anyone who, in a field of, say, 6 hectares stocked with 25 ewes and twins per hectare, has tried to take out 30 ewes and be sure of having the right lambs with them, will know what I mean. Yet some changes may well be necessary, and there are two ways of achieving it without too much fuss. The first is to err on the side of under-stocking and take up the slack, when it occurs, by fencing off a strip of the field, now easily done with plastic electric netting. The fenced-off strip can be added to the conservation area, and this is really only easily done when the conservation is in the form of silage. Or, secondly, to use the ewe hogg replacement flock – that is, the dry ewe hoggs being kept on for gimmer replacements – as the means of adding to or substracting from the stocking rate. They at any rate can be easily separated from the ewes and lambs.

All this sounds fine and good on paper, but what about the realities of life – unpredictable weather and all the other unforeseen snags that come up year by year, even in the most well-ordered and managed of farms. Is the sort of stocking rate which I am talking about during this period of May to July a feasible proposition? Of course, if you have bad land, or worse still, if you have poor quality pastures – the reverse of the thick carpet principle – then you would be in trouble. But given only reasonable land and good, dense pasture – and it is surprising what wonderful pasture can be grown even on bad land – then the answer must be, yes, it most definitely is on.

In my early days, I got a lot of inspiration from the Romney Marsh. An area of marvellous quality permanent pasture on admittedly high quality land. The productivity of the pasture had been created over the generations by very heavy stocking rates indeed. The traditional management system on the Romney Marsh was, and to some extent still is, very special, but what is for sure is that it supported numbers of sheep that make my 25 ewes and lambs per hectare look modest indeed.

After shearing, the fields in the Marsh look as if they are covered with driven snow, they are so white with sheep. The dictum there was, that if the field was properly stocked, you should be able to throw what is now a 10p piece as far as you could and be able to find it immediately without any difficulty. If you could not walk straight to it, the field was under-stocked! The grass, in a dense carpet, growing to the sheep. It looks as if there is not enough for them to eat; but take the flock out for 24 hours and you will then see just how much grass has grown during that short period. And that really is the essence of high stocking rates at set stocking.

MIXED GRAZING

Finally, some thoughts on mixed grazing of cattle and sheep.

Traditional thinking has always been that this was the right way to achieve the very best results from both species. Their grazing habits were complementary and one diluted the stocking rate of the other leading to less competition, and, in particular, fewer worm problems. It is my belief that this reasoning is only acceptable and workable at low overall stocking densities. It works because the pressures in total are not great, and nor therefore are the overall returns. Judged on the performance of an individual lamb or bullock, it looks fine. Judged on output per acre, it just is not feasible at present day land values. As soon as you push the intensity up, the system falls apart due to serious worm problems – and, of course, this is especially true on long-term or permanent pastures.

Recent work at the East of Scotland College of Agriculture has led the way towards intensive production of both cattle and sheep without these worm problems. It is based upon the generalisation, that is broadly true, that sheep and cattle are not affected by the same internal parasites, and that a sufficiently long period of time without either cattle or sheep will clean up a pasture of their specific worms simply because the life cycle is broken. The system developed from this fact is based on a three-year rotation, each separate year in turn being devoted to only sheep, then cattle and then conservation. There is sufficient practical experience of this system now to say that for long-term or permanent pastures, it is the key to achieving high output. I would seriously advise anyone whose farming includes this type of pasture – which outside the arable areas must be the majority – to study this system in depth and apply its principles to their own operation. Seven farms recorded by the East of Scotland College who adopted this system showed the following average results which speak for themselves.

Table 8. Results of rotating sheep, cattle and conservation

	1974	*1977*	*% increase*
Average farm size	221·5 ha	221·5 ha	—
Area of cereals	41·3 ha	51·0 ha	23·5
No. of ewes	420	508	21·0
No. of cows	87	99·5	14·5

WINTER STRAW FEEDING AND THE EFFECT ON GRASSLAND MANAGEMENT

I have described in the previous chapter how treated straw may be used very satisfactorily as the basis of the winter's bulk feed. The implication of this is that it is no longer necessary to make hay or silage as the conserved feed for the winter. So we have to stop and think just what effect this is going to have on our grassland management and the stocking rates that we can achieve.

Let us go back to the curve of grass growth in its exaggerated form in Fig. 4. Against that curve, the line representing the flock requirements for food shows up periods of both surplus and shortage. Traditionally we have dealt with the surplus by conservation into hay or silage, and in the next chapter I shall be describing more fully how we can fill in the summer gap with roots.

What will happen if we keep the stocking rate the same and do not make hay or silage? When the flush of growth comes, the grass will get out of control due to undergrazing. The consequences of that are serious in that nutritional quality and palatability will fall off rapidly. That is bad enough in the short term in that it will affect growth, but worse still it will reduce the productivity of the grass right through the remainder of the season. So clearly this is an option which we cannot accept.

Should we then increase the stocking rate so that there are sufficient ewes and lambs to keep control of grass production when it is at its peak? That would certainly solve the grass management problem but equally certainly it will aggravate the problems of under supply not only in the summer gap but also in the critically important period in late March and April. That too is surely out of the question, unless we do something to fill in these gaps. And we come back to roots.

High stocking rates are such an important element in maximising profitability that it really is worth all the scheming and effort that may be required to push it up to the top – always providing that lamb growth and quality are not sacrificed along the way. The question to be asked is – can we add another three or four ewes to the hectare if we are not obliged to set aside land for hay or silage? And the answer is yes, certainly, and successfully too, so long as we grow sufficient roots to fill the gaps. Mangolds or swede stored to feed at grass during March and April until the flush comes. Turnips – either white or fodder – grown as a crop to be folded by the weaned lambs in July and August.

Once you have accepted the necessity of being a root feeder of sheep as well as a grazer – and the absolute need to do this in a planned and professional way – then all of a sudden you begin to realise just how attractive straw can be. Obviously it only makes sense in arable areas where good quality barley straw is available just for the baling. Under these conditions, consider the advantages:

(1) Use is made of a by-product which otherwise is a disposal problem. Even if straw burning is not banned, we surely cannot continue to burn it near houses and villages.
(2) All the extra work of hay or silage making is done away with – to say nothing of the need for expensive equipment, especially for silage making.
(3) Straw is a product of reasonably consistent quality. Even the most dedicated hay maker can hardly claim, in our climate, to be sure that he can always produce the best.
(4) Most important of all – it takes the pressure off the grazing in May, June and July. To make sufficient winter food from grass means shutting up anything up to 40 per cent of the grass area with effective stocking rates going up from 15 ewes and their lambs to somewhere around 22 or 23 per ha. Run into drought then and you have a real problem on your hands.

All this adds up to a strong argument in favour of straw. We are using it now at Worlaby and whilst nothing is ever completely straightforward, I think I can say with some confidence that we shall not be going back to haymaking. Most of all, I appreciate the relaxation of the management pressures on sheep and grass during the summer. We are free to go for higher stocking rates, at the same time being confident that the sheep are doing well – and without having to look over our shoulders to see if we are going to have enough food for the winter.

Chapter 10

FODDER CROPS

AT SEVERAL points in this book I have laid considerable emphasis on the importance of resolving the problems caused by the 'summer gap'. This period in July, August and September, when the curve of grass production dips down into a trough, coincides with the time when total liveweight per hectare is probably at its highest. Shortage of food then will lead to stunted store lambs, and ewes which will be in a poor condition for going to the ram later on. Except with the early lambing flocks where lambs will have been sold before July, the resolution of this problem is an essential element in the success of intensive sheep production.

Fortunately, the solution is not difficult and lies with the use of root crops. For those of us who are also arable farmers, this presents no problem from the point of view of equipment, particularly if we are also sugar-beet growers. It would seem, at first sight to be rather more difficult for the all-grass farmer to justify what seems to be such a radical change in his farming methods and very probably on land which is not suited to cultivation. Again, fortunately, the solution is relatively simple and easy to adopt. For we have seen a real breakthrough with the introduction of direct drilling, allied to both root growing and the renewing of pastures. On top of this have come the new varieties of quick-growing forage turnips which can be so drilled that they arrive at maturity when required.

The way is open to solving the problem of the 'summer gap'; but the prospects are much wider than that as we can now think of using root crops at any time of the year when it is appropriate and they fit in to the overall scheme of intensifying sheep output. It is an important step in the adaptation of the root folding technique of the last century to modern conditions.

WHICH CROPS?

Of course, it all depends on what time of year we are aiming at. Some crops are quick growing, some are slow, some are winter-hardy, others not. Given below is a necessarily brief summary of those available. For more detailed information on varieties, yields and growing techniques, the reader would be well advised to consult ADAS.

Forage Turnips

Sometimes known as Dutch stubble or 12-week turnips, they were introduced in the mid-60s and there are a considerable number of varieties to choose from. Ponda, Vobra, Taronda and Civasto are all useful, and others will no doubt follow as development is going on all the time. The essential feature of this most useful plant is that it gives a high yield of nutritious fodder very quickly indeed, somewhere around 10 to 12 weeks from date of sowing, depending upon the time of year. Much of the yield comes from the leaf, which in any case seems to be more palatable than the root.

The forage turnip is my choice for the summer gap. We aim to sow the first drilling by April 15th and this will give us a full crop onto which we wean lambs on July 1st. It is not worth drilling earlier than this, certainly in eastern England, because of the risk of frost which will cause the turnips to run to seed. Once this has happened, then much of the food value has been lost. I have to admit that it is still a risk when drilling in mid-April, but one worth taking nonetheless. Then subsequently, we follow up with staggered drillings every two weeks until the end of May. By this means we are assured of a succession of food from the first week in July until the end of September. Indeed there is no reason why the forage turnip should not be used to provide food until December in an average year. It is not winter-hardy, however, and there are other crops which are much higher yielding. As to yield, obviously this can vary greatly according to the success with which it has been grown, but as a rule of thumb, I calculate on a stocking rate of 60–70 lambs per hectare over an eight-week period.

Giant or Forage Rape

Rape was the traditional crop to be sown for autumn feeding of lambs, particularly as a catch crop. It is not quite as quick growing as the stubble turnip and relies for its yield entirely on

20. Kale direct-drilled into old pasture burnt off with Gramoxone.
Source: David Lee.

21. A heavy crop of Ponda forage turnips grown as a catch crop after vinir peas.
Source: Author.

its leaf. It still is a very popular choice, particularly in the North where its rather greater degree of frost resistance is an attraction. Where it has been quickly grown and is 'lush', there can be a danger of rape poisoning. Not too much seems to be known about the causes of this problem, which can result in deaths as well as poor growth sometimes associated with skin symptoms. Suffice it to say that lambs should be introduced slowly on to rape until they are fully accustomed to it, and also that the rape plant should be relatively mature before use. Yields are similar to the stubble turnip.

Fodder Radish

A very quick growing plant which is really only of any use as a catch crop. Its great disadvantage is that it runs to seed at the slightest excuse and is then of low digestibility and almost completely unpalatable. In my view, it is not worth considering (except perhaps as pheasant cover!).

Turnips

These are what I call the old-fashioned turnips, either white or yellow fleshed, and which, unlike stubble turnips, depend for yield almost entirely on their roots. Relatively fast growing, they can be sown in the spring for summer use or after harvest as a catch crop for late autumn use when they are relatively hardy. They do have the advantage over stubble turnips that it is the root which provides the food and thus they will 'hold' in good condition for longer, particularly in hot weather. Perhaps the real advantage, under the right circumstances, is that they can be dual purpose. In Lincolnshire, for instance, they are often grown as a catch crop after vining peas or early potatoes, and the roots harvested for sale to the processing companies for inclusion in their mixed vegetable packs, the residue being folded by sheep.

Swedes

Whenever I go north, I am always filled with envy at the efficiency with which the Scots grow swedes. At its best, a well-grown crop of swedes gives a high yield of up to 90–100 tonnes per hectare. Probably the folding crop par excellence for fattening lambs right through the winter. And incidentally, they are now often harvested mechanically and stored and then subsequently fed, chopped or whole, to cattle or sheep in yards.

My envy used to lead to frustration – we could not in Lincolnshire grow these sort of high-yielding crops. The major problem of course was mildew which can be an absolute killer for the swede. However, modern technology has come to our rescue with the sulphur-based spray which really does the trick. We shall never get the same high yields as in the north, largely because of drought, but we can at least get respectable yields. And in my view, the swede, harvested in November and stored, is probably the best crop to supplement the grazing in April – the pre-grass flush gap which if not filled can undo all the good you have done with careful pre-lamb feeding. A field of hungry ewes with young lambs in March and April, with milk yield disappearing in front of your eyes, is not a pretty sight.

Mangolds and Fodder Beet

Mangolds have the virtue of being potentially the highest yielding fodder root that we can grow. One hundred and twenty five tonnes per hectare is by no means impossible on good land. They are not, however, winter-hardy and have to be lifted before winter sets in. That in itself would not present any great problem because they store well if adequately protected in the clamp. The real difficulty lies in the harvesting. They are of an awkward shape, which does not lend itself readily to the harvester, and they are tender and bruise easily. Once bruised or broken, they can deteriorate in store. Hand harvesting is really the only solution, which to my mind puts them out of court. This is a pity because they are a very convenient food for use in March/April for recently lambed ewes.

When grass is in short supply, and it is so essential to keep the milk supply going, a supply of mangolds has great value. Sheep obviously like them and scattered over the field they provide a convenient supplement, which avoids the stress and panic which ensues when once-a-day hand-feeding of concentrates is necessary. The daily rush to the troughs is an upset and a cause of mis-mothering which should be avoided at all costs, short of compromising the milk yield.

However, there is the fodder beet, which I suspect has been a neglected alternative. Not as high yielding, in terms of gross root yield, as the mangold, but higher in dry matter. But it can be harvested mechanically either with sugar-beet or swede harvesters. Personally, I think that we ought to be growing the fodder beet as a 'security feed' to cover the risks of a cold and backward spring.

Cabbages and Kale

A subject for a book by itself, so I must resist the temptation! The kales, marrow stem and thousand head, are well known, particularly in the dairy world, and make a good winter forage crop for sheep. But, although giving relatively high yields, around 50 tonnes per hectare, it always seems to me that a considerable proportion of this is wasted by sheep. The leaves, in which is found the bulk of the value, are easily broken off and trodden under foot, and I never go into a kale field being grazed by sheep without being appalled at the amount of wasted leaf lying on the ground. To my mind, this is where the cabbage has a great advantage. The leaf is, of course, tightly bound around the head or heart of the cabbage and cannot be broken off, and the whole plant is solid. It is quite remarkable to watch a flock of lambs eating cabbage. They will stand, in twos and threes, around a cabbage, eating away at it rather as one may munch at an apple. Very impressive! And as a bonus, properly grown – that is, sown really early and fertilised heavily and using the right varieties – the yield can be tremendous. One hundred to 125 tonnes of green matter per hectare is quite possible, with a dry matter yield of 12–18 tonnes/ha.

There is really nothing to touch the cabbage for a winter food for lambs, certainly in those areas where swedes cannot be grown. Much of the development work on cabbages as a forage crop has been carried out by ADAS at Beverley, Yorkshire, under the inspiration of the then vegetable specialist, Fred Tyler. That is the right place to go to, to seek detailed information.

Arable By-products

No discussion of fodder crops would be complete without a mention of the great potential for sheep profit that is there for the taking in the use of such by-products as sugar-beet tops,

Brussels sprout waste, and commercial cabbage and cauliflower residues. Not everyone lives and farms in an area where such crops can be grown, but for those who do, and have good commercial reasons for growing these crops, then there is a lot of good food there for free.

Sugar-beet tops are, I suppose, the most important. Certainly an excellent food for lamb fattening and coming as they do, in the period of October to December or even January, they fill a need that could only be otherwise provided by a specifically grown crop. Thirty-seven to fifty lambs per hectare over a four-week period, provided the tops are well saved and not too soiled, is a useful bonus on top of the sales of beet. Why is it, then, that no more than some 20 per cent of beet tops are used for stock feeding? Mostly, I suppose, because arable farmers prefer to avoid livestock complications and know their own limitations. A desire to get the land quickly into winter wheat certainly plays a part. Also many modern harvesters smash the tops up so that they rot down and disintegrate very quickly. But the higher profitability of sheep over the past few years, and in particular the price incentive to keep lambs on for slaughter in mid-winter, really ought to make us think again. With the need to intensify production to the maximum in face of rapidly rising fixed costs, the waste of such a valuable by-product is difficult to justify.

THE PLACE FOR ROOT CROPS IN A SHEEP SYSTEM

From what I have said, all too briefly, about the different crops available, it is evident that we have the choice that could give us fodder from early July right through to the following April. The convenience of this is that it ties in so well with our main fodder crop, grass, which is at its peak production from April to July.

The point that I want to emphasise is that, in planning the intensification of a flock of ewes, we should not fix our minds solely on grass. To do so is to impose limitations on output that are severe both in terms of stocking rate per hectare and in quality of lambs produced. The remedy lies with a strategic exploitation of the various root crops to provide adequate nutrition throughout the 12 months of the year; and to do this in order to complement the equally maximum exploitation of grass.

(The following Tables 9 and 10 set out the availability of the different crops throughout the seasons.)

Table 9. Details of Forage Crops Available for Summer Feeding of Lambs

Type	*Season of use*	*Yield fresh weight per hectare (tonnes)*	*Dry matter yield per hectare (tonnes)*	*Days feeding for lambs*
Continental/ stubble turnips	July–September	86 (70–99)*	7·7	99 lambs/ha for 6–8 weeks (4,900 lamb feeding days/ha)
Fodder radish	August–October	37 (25–44)	5·2	50–60 lambs/ha for 6 weeks (2,350 lamb feeding days/ha)
Kale	August–October	44 (29–50)	6·2	60–70 lambs/ha for 6 weeks
Forage rape	July–October	32 (17–40)	4·5	60 lambs/ha for 6–8 weeks (2,700 lamb feeding days/ha)

*Typical range

Table 10. Details of Forage Crops Available for Autumn and Winter Feeding of Lambs

Type	*Season of use*	*Yield fresh weight per hectare (tonnes)*	*Dry matter yield per hectare (tonnes)*	*Days feeding for lambs*
Continental/ stubble turnips	September–December	37 (30–60)*	3·4	60 lambs/ha for 6 weeks (2,500 lamb feeding days/ha)
Forage rape	September–November	28 (16–37)	3·9	50–60 lambs/ha for 6 weeks (2,300 lamb feeding days/ha)
Sugar beet tops	October–November	17 (9–24)	2·7	35–40 lambs/ha for 6–8 weeks (1,800 lamb feeding days/ha)
White fleshed turnips	October–December	66 (50–85)	5·9	50–60 lambs/ha for 6–8 weeks (2,700 lamb feeding days/ha)
Kale	September–November	35 (25–40)	4·9	50–60 lambs/ha for 8 weeks (2,300 lamb feeding days/ha)
Yellow fleshed turnips	November–January	60 (44–74)	6·0	50–60 lambs/ha for 6–8 weeks (2,700 lamb feeding days/ha)
Winter cabbage	November–January	85 (61–98)	12·7	70–80 lambs/ha for 8 weeks (4,200 lamb feeding days/ha)
Swedes	November–March	60 (44–74)	7·2	75–95 lambs/ha for 12–14 weeks (7,700 lamb feeding days/ha)

*Typical range

Source: MLC Date Summaries of Upland and Lowland Sheep Production, 1978.

In fact, it will be the lambs which will be the major consumers of the roots. The pattern of grass growth and its integration with conservation, be it hay or silage, fits in well with the needs of the ewe flock. It is for the lambs whose needs come to a peak in July after weaning and carry on into autumn and perhaps winter, depending upon the farm and its system, where the root crop has its real value.

At the risk of sounding like a computer, the food requirements of the store and the fattening lamb flocks can be programmed and provided with considerable precision. This, remember, is exactly what the old arable folding system did with great success. There is no reason why we should not follow exactly the same principle. On the arable farm, in particular, the scope is enormous, especially for the exploitation of catch cropping.

Now that winter barley has become so firmly established, the opportunities are increased for drilling a crop of stubble turnips into the early harvested stubble. Of course, if you are, like me, a grower of vining peas, or, unlike me, of early potatoes, the opportunities are even greater. But, and I emphasise this, it demands a determined and professional approach, where the farmer believes in the worthwhileness of what he is doing. A slap-happy, next-week-will-do approach will lead to disappointment and deservedly so. For days count, and the difference in yield of forage turnips sown between, say, the first and second weeks of August is very noticeable. And perhaps, even more important, moisture conservation is the essence of success in getting quick and vigorous germination – which is one of the reasons why direct drilling is so attractive.

The all-grass farm also has its opportunities and again these are associated with direct drilling. I see every reason in favour of incorporating a planned area of roots into an all-grass farm, not only for its own value for sheep feeding, but also as a component of pasture renewal.

DIRECT DRILLING

You will, no doubt, have already picked up the point that I am an enthusiast for this technique. Not, I hope, to the extent of being blind to the need for appreciating its limitations. Nevertheless, I reckon it to be one of the major breakthroughs for the livestock farmer that have come his way in recent years. Although this is a book which is primarily about sheep, the direct drilling technique is so important that it merits some fairly detailed examination.

Why not plough and cultivate? There is no special virtue in ploughing *per se* except as a means of burying and destroying surface residues, be they the grass sward or trash and rubbish. Ploughing has two great disadvantages, apart from its cost. Firstly, it loosens the soil and thereby loses moisture: secondly, whilst burying the trash, it also buries the top few inches of fertile soil and brings to the surface the much hungrier subsoil.

Moisture conservation, often an essential in catch cropping, and retention of fertility in the immediate top soil are two of the great advantages of direct drilling. Furthermore, there are many cases in the grassland areas where ploughing, or any other land working, is out of the question; either due to rock, or slope, or wet areas.

How, then, to kill the surface sward or trash? The other essential component of the system, in addition to the drill itself, is the use of Gramoxone (paraquat) to burn off before drilling. This marvellous chemical, which destroys everything green and growing on the surface is itself destroyed on contact with the soil so that there is no residue, does not, however, kill plants with strong roots such as couch grass, thistles, docks, etc. For these much more difficult weeds, materials such as Roundup and Asulox have to be used.

THE PLACE FOR DIRECT DRILLING

Really, there are two places. Firstly, the arable context, for root drilling at any time of the year, but of particular advantage for catch cropping behind the arable crops, usually cereals. The saving of both vital time and moisture are greatly in its favour. Of course, the drill as a piece of equipment can be used as a normal drill, both for cereals and grasses as well as roots and equally in a cultivation situation as for direct into unmoved ground.

On an arable farm there is a great deal to be said for drilling new leys in the autumn. These can follow either wheat or barley, although winter barley has much to commend it, coming so much earlier to harvest. Drilling into stubble on heavy land is much to be preferred to undersowing the cereal crop in the spring. For one thing, all the evidence shows a significant reduction in cereal yield if undersown, especially when you have got more than one eye on wanting to establish a good ley.

Of greater significance, perhaps, is the fact that we can now spray our barley against wild oats, whereas when undersown this wasn't possible. It was infuriating to spend a lot of money in controlling this horrid weed in our other crops in the rotation, only to see all the ground gained duly lost during the barley year. Incidentally, we also get a much better take of grass seeds now that we are bare sowing in the late summer. No more of those annoying bare patches where the undersown grasses were killed out by laid patches in the barley; bare patches which were an invitation to an invasion by couch grass. All in all, a formidable list of advantages for the direct drilling technique on the arable farm.

Secondly, in the context of the grass farm. Whereas on the arable farm one can manage perfectly well without the direct drill, it is merely a technique which has many advantages; on the grass farm, I dare to say that it is, to all intents and purposes, essential.

On another farm, for which I am responsible, we have the problem of a significant proportion of the total which has to remain in grass. These areas, which are of the poorer soil types anyway, have springs which appear in all sorts of odd places as well as rocky outcrops. Consequently, the land is totally unsuited for cultivation and even if it could be cultivated, would not produce any crops worth having. On top of all that, the permanent pasture when we took the farm was really of abysmal quality. This description could be applied to many tens of thousands of acres throughout Britain. How to convert it back to productive permanent pasture without high levels of totally unjustifiable expenditure, should be the question any occupier of such land should be asking himself.

The solution we have used is based on the concept of 'pioneer cropping'. First of all, in the preceding autumn, the appropriate dose of both lime and slag following soil analysis. In the spring, the destruction of the sward with Gramoxone, and it is very important to follow the ICI recommendations to the letter for the double dose programme. That is to say, a second spray of Gramoxone 10 days to a fortnight after the first. It is, we have found, completely false economy to miss out this second application which gives the *coup de grace* to any resurgence of growth from the thick mat of the old pasture. Incidentally, and of considerable importance, this killing process is greatly helped if the pasture is tightly grazed beforehand. Immediately after this second dose, in goes the direct drill to sow the pioneer crop, with a top-dressing at the same time of 60 kg of N per hectare.

As a pioneer crop we have used, with equal success, forage turnips either on their own or cross-drilled with Italian rye-grass. On the whole I prefer the latter, as the Italian can stay down for two years. During these two years, nitrogen top-dressing should be generous and the pasture should be grazed really hard. The objective is to stimulate biological life in soil which is often 'dead' and asphyxiated, to restart a fertility cycle of rotting down raw organic matter via grazing pressure and released nitrogen.

Two years of this treatment and we expect to go back into a long-term grass mixture. This pasture can then either stay down for as long as its quality is maintained or, as I prefer, on a five-year rotation, be put into a forage turnip break in the spring, to be resown again that autumn. That way we have one-fifth of our 'permanent' grass acreage in April – and May-sown turnips each year, and our pastures in a continuing state of renewal. And it brings us back to provision of food for the summer gap.

I have cited our experience on this farm at some length because it seems to me to illustrate so well the potential that the direct drill can introduce on to a farm where the conditions are far from easy. From this description, it should be apparent that, given a degree of imagination and a determination to exploit the potential, there really are very many circumstances where this idea can be applied with great benefit.

DISADVANTAGES OF DIRECT DRILLING

After all that, I ought to introduce an element of balance! There are, of course, some snags – the capital cost of the drill being not the least of these. For that reason, and also because the output of the drill is often more than one farm needs, it does seem to me to be a machine well suited to co-operative ownership. And why not?

Other snags? Not perhaps so much snags as a realisation of its limitations. It is just no good at all direct drilling into couch. You have got to kill it beforehand, and the means to do so are therein the form of Roundup (Glyphosate). It is an expensive cure, but there it is, you really have not got any choice.

Then there is the question of soil conditions. You must avoid leaving a 'smeared' slit. It is really common sense that you cannot expect any seed to grow in that sort of environment where either water cannot get away or, alternatively, the slit dries out like concrete. Of course, this problem is worse on clay than on lighter soils, and with some makes of drill than with others. And on such soil conditions, you must avoid badly compacted land. The sort of thing that can happen after a wet harvest with heavy trailers having run about all over the surface. In fact, I believe that the subsoiler is a very desirable part of one's weaponry in the art and practice of soil management.

Finally, the need for drainage. Nothing will grow if it has its feet in water, and direct drilling is no more a short cut to an easy life than anything else. However, the attraction of the direct drill where you have limited areas of wet, is that you can work your way around them and improve perhaps, say, 80 per cent of the field. This may well be a very acceptable alternative to expensive drainage on land which is not good enough to justify it.

THE UTILISATION OF THE FODDER CROP

Almost synonymous with roots for sheep is the concept of 'folding'. It goes back to our impression of the traditional flock of folded sheep on the arable farm – a concept which has given me so much inspiration. But here is something which demands some serious questioning. No one can afford to contemplate the moving of a daily fold, however desirable that might be in any evaluation of liveweight gain.

Fortunately, there is no need to consider any such thing. The invention of the electrified plastic fencing net has changed all that. Light and easy to move in a matter of minutes, folding, albeit for a more practical three or four day move, is now once more a practical proposition.

More questioning, however, is warranted. Is there any real justification for folding at all for such 'leafy' crops as forage turnips or rape or kale? I begin very much to doubt it. I prefer to give a bunch of lambs the run of a whole field, though perhaps it is as well if the field is no more than four or five hectares. They then settle down quietly in the field, and you will find that they effectively graze from the edges anyway. Contrast this with the state of excitement each time a fold is moved. The present strip is virtually eaten out and the lambs will have been on pretty short rations during the preceding 24 hours. As soon as they are let into the fresh fold, they will dash in like kids at a tea party. Not only do they gorge themselves, but in the process trample on the crop and waste much of it. I return to my sentiments expressed in relation to forward creep grazing. I cannot see that such unnecessarily induced excitement and stress is anything other than positively damaging to the performance of any livestock. And it is a happy coincidence that set stocking of fodder crops also reduces work.

Chapter 11

SHEPHERDS AND SHEPHERDING

When all the planning and the thinking has been done, all the advice taken, the success of any livestock unit will depend totally on the man who is actually carrying it out. The same thing, of course, is true with the growing of crops, although to a lesser degree, and that should be no surprise to us as we are dealing with living things, plants and animals, that do not conform as individuals to a pre-determined factory plan. Thank goodness!

STOCKMANSHIP

What is stockmanship? Not something that can be taught in a classroom or learnt from reading a book, that is for sure.

It is attention to detail; the day-to-day attention to the small things that if they are not right have an insidious effect on overall performance; the precise and prudent use of all the tools which modern science has given us.

It is an acutely developed sense of observation. It is so easy to walk past something without having seen it. Most people see what they want to see. Familiarity not only breeds contempt, it also affects the eyesight. So it is observation – continually asking questions – and the ability to look through the mass and see the individual.

It goes without saying that it is technical skill.

It is also what I call an acutely developed sense of smell. The ability to smell trouble before it happens; a sort of sixth sense which anticipates the problems long before they are evident to lesser mortals. All good stockmen have it. I wish I knew how to include it in a training programme.

It is, and let us not be afraid of appearing to be sentimental, a love for animals and for the individual. A sensitivity which can withstand the pressures of economic reality.

THE DILEMMA

We live in an era when the employing of people is hedged around with difficulties, rules and regulations. If employing them is difficult, getting rid of them if they are not up to the job – to my mind one of the most distasteful things any humane employer is faced with – is very nearly impossible.

On top of these legislative complications, we live in an economic environment where agriculture is under pressure, and in a social environment where everyone has a right to more and more of the 'good things' of life and more time in which to enjoy them.

It is not the purpose of this book to argue for or against the reasoning that lies behind this state of affairs. Merely to recognise the fact. And at the same time we must recognise that we cannot afford the luxury of a shepherd looking after 400 ewes; in other words, a man of leisure who has the time to contemplate and to ponder and to notice.

Herein lies the dilemma. How can we reconcile the definition of stockmanship which I have attempted to give you? How can we reconcile the qualities that it demands from a shepherd with the pressure of being responsible for at least 800 ewes – from which we expect not 130 per cent productivity but 175 per cent minimum? And, moreover, achieve this without straying too far into what are called 'unsocial hours'. I do not pretend to have any magic formula, nor even to have solved the problem, if indeed it is solvable. Of one thing I am certain: we have to scheme to the limit of our ingenuity to try and provide the kind of environment where stockmanship and mass production can to some extent go hand in hand.

HOW TO EQUIP YOUR SHEPHERD

One thing that we can do is to design buildings, and plan mechanisation, so as to reduce to the absolute minimum the time and effort spent on manual work. Too often such investment is judged on whether it actually reduces labour costs. There are many times when it can do just that, but perhaps the real criterion ought to be the extent to which it frees the stockman and enables him to practise his art.

The Buildings

I have already written at length on the advantages of inwintering, and they apply greatly in this context. Within the building there is much to be done, in order to introduce convenience. From the big things, such as the arrangement of doors and pens so that the building can be quickly mucked out with the appropriate machinery, and so that feeding equipment can get to every pen; to the small details, such as plenty of lights and power points conveniently situated; water which does not freeze up; and a properly equipped store and basic office.

Machinery

The great advantage of silage over hay is that it can be fully mechanised from start to finish. The complete diet feeding line of thinking developed for dairy cows can be applied to sheep feeding. Concentrates can be mixed into the silage via a dispenser fitted to the forage trailer. The danger is that feeding will become an unthinking 'canteen'-type of exercise. But if it does, that is rank bad management and no excuse for not making use of machinery.

Fencing

There is no greater frustration for a good shepherd than not being able to control his sheep and their grazing because the fencing is either not good enough or is non-existent. Temporary fencing, good as the electrified nets are, is no substitute for, at the very least, good boundary fencing around each field.

Handling Equipment

Properly designed pens and dips are surely an essential.

HOW TO PAY YOUR SHEPHERD

You cannot buy stockmanship, and doubling a shepherd's wage certainly will not make him twice as good a shepherd. That is a blinding statement of the obvious if ever there was one! Equally, no matter how important wages are as an element in farm costs, you cannot judge a good man to the last

£100. The effect that he can have on financial output far outweights any savings that might come from bargaining hard on salary fixing. But, I repeat, money of itself does not ensure success.

Shepherds are, however, human like the rest of us, and even the most dedicated amongst them respond to financial incentive. How then to do it?

I think most people would agree with me that time sheets and hour-keeping are completely out of place in a stock job. So a flat wage for the job must be the right start, and it has to be at such a level that it gives security and financial well-being month by month. It has also, of course, to be competitive with comparable employment in the area, for it will be this basic wage that he will talk about when he goes to the pub, not his overall income at the end of the year after bonuses. Incidentally, I happen to think that it should also include a good pension and life insurance scheme as well.

On top of that, however, I believe that we need an incentive scheme pitched at such a level that it actually hurts his pocket when things go wrong and rewards him at an appreciable level when they go right. The definition of what it is that is going right or wrong must be that which affects the profitability of the flock. It is for this reason that I am not the least bit interested in paying a bonus on lambs born, or even lambs weaned. A lamb is of no value to me, and represents no profit, until the day it is sold. So the right place to put a production bonus is on lambs sold, and this is what we do at Worlaby. Each month the number of lambs sold (or with ewe lambs, retained in the flock for breeding) are noted and paid for per head. This concentrates the attention most wonderfully, and there is almost equal lamentation amongst all of us if a lamb should die a week before it is due to go for slaughter.

The snag in this reasoning lies with the weather, as in so many things in farming. A shepherd can be getting his lowest bonus in those years when, due to bad weather, he has had to put the very maximum of effort into his job. And, conversely, can be getting his biggest reward following the easiest of lambing seasons. In a way it is a snag, but in another sense it associates him very fully with the real life-blood of the business. The fact remains, however, that he will be penalised financially as well as being totally disheartened in the really bad lambing season. How to get over that is part of the art of good bossmanship – which is about as complex and impressive an art as is stockmanship.

HELP AT LAMBING

We all accept that maximum economy in the use of labour, consistent with high production, has to be one of our major preoccupations. Having said that, lambing is most definitely not the time to be cheese-paring with labour. It must be false economy to be short-handed during the critical six weeks of lambing, and, of course, the overall work load is way out of proportion to that during the rest of the year.

Such a lot can be done to save those extra lambs which will make such a difference to the flock profit. In our experience there is no better way of doing this than by employing women. No doubt I shall be accused of contravening all sorts of stupid legislation which pretends that no distinction must be made between the sexes. As the French say, 'Vive la difference', and I know that a woman's maternal instincts fit her exceptionally well for this job of lamb preservation.

In Scotland, it has for long been traditional practice to employ a so-called 'lambing man' to help the shepherd. It is the lambing man who actually does the lambing and the sitting up at night, and the shepherd fetches and carries and provides. A variation of this on a farm in Wiltshire with which I am connected is that there they can use, for many jobs throughout the year, but particularly for lambing, the services of a contracting shepherding service. Very efficient they are too, albeit rather expensive. But the cost has to be judged in the context of the fact that the shepherd is released so that he can and does look after nearly 3,000 ewes.

Whatever the system, there is no doubt that extra help at lambing is well justified.

HOW TO FIND YOUR SHEPHERD

The old saying that 'shepherds are born and not made' must be regarded as a counsel of despair, by agricultural colleges and

Agricultural Training Board alike. Nevertheless, good shepherds are in grieviously short supply, and I do not think that is really anything to do with wages. Supply and demand has sorted that out reasonably well.

Colleges such as Kirkley Hall in Northumberland have run a specialist sheep course for many years now. I have every reason to think that it is an extremely good course, but for many years the majority of students that it attracted were farmers' sons destined to go back home. Happily however there are now signs that that is changing and Kirkley Hall, together with several other of our County Colleges, is turning out good potential shepherds.

Is a college, however good, the right place to start training a future shepherd? In these days when a knowledge of technical complexities such as disease control and vaccination is such a necessary part of a shepherd's equipment, it certainly ought to be. But it does not seem to be working out that way.

Should the recruitment and training be based upon the Agricultural Training Board via its Apprenticeship Schemes, and then subsequently through shepherds' discussion and in-job training groups? I think the answer must be a positive 'yes', and as an industry we can be criticised for not putting much more muscle behind this. Sadly, however, in farming, where there is a declining labour force anyway, and in economic circumstances which are often far from easy, there is some considerable difficulty in fitting in a young man as an extra.

Traditionally, we have relied upon the advertisement as our means of recruitment when the need arises. It still remains the basic way of finding and attracting the man when you need him. And I am bound to say that my own personal attempts to teach young apprentices have not exactly been crowned with success. Nevertheless, it is not a situation which is either constructive or enlightened.

Chapter 12

SHEEP HANDLING

NO MATTER what the size of the flock, some form of handling equipment so that the sheep can be collected together and treated is essential. Essential not only because it enables the carrying out of such jobs as drenching, vaccinating and foot-rotting to be done quickly and efficiently, but also because of the legal requirements for dipping against sheep scab. That is a bald statement of the fact of the situation. From there on, very many variables come into the picture following individual preferences.

HANDLING PENS

When we started our flock at Worlaby back in 1960, one of the first things I did was to put in a combined set of handling pens and dip designed to cope with all the work involved with 1,200 ewes. At the time it cost a lot of money, although it does not seem very much by today's standards, and it is interesting to look back and reflect how very differently I would have done it today.

The basic idea was that we would have a set of fixed pens, of oak posts and rails, set in and on a concrete base, and that it would be positioned centrally on the farm so as to minimise sheep movements. Of course, with a farm as large as Worlaby, this still involved moving sheep considerable distances. This is a very definite disadvantage, not only because it wastes time, but also, particularly in hot weather, can cause serious stress to the sheep. So although the principles of the original design still hold good, I would make two fundamental changes were I to be starting afresh today.

Firstly, I would not make the pens fixed and permanent. Rather I would prefer to put down a number of strategically sited concrete slabs in different areas of the farm so that movement distances really were cut to a minimum. On any one of these areas, all handling operations could be carried out with the exception of dipping. But because the dip really has to be a permanent installation, and one could hardly justify the cost of more than one, one site would have to be chosen for it.

The second change follows from the first. Instead of putting down permanent post and rail fences, I would use a set of movable metal hurdles (see photo 24). The great advantage of these is that they are portable and can be taken to the sheep rather than the sheep to the pens. They are also much more adaptable, and pen sizes can be changed to accommodate different sized lots of sheep. Incidentally, but of no little practical importance, the same hurdles can be used in the wintering shed for the winter work.

Basic Principles

Let us, however, return to the basic principles which I believe are valid whatever the size of unit or type of flock or farm.

(1) The pens should be of differing sizes. A holding pen which is big enough to hold all of the biggest lot of sheep that is likely to be brought in at any one time. This should lead in to a number of progressively smaller pens so that different lots can be handled. Sheep *must* be tightly packed in a pen for ease of handling. If they can keep running around you, both you and your sheep will suffer.

(2) The flow from one pen to another should be smooth. It should follow a natural circuit, preferably 'funnelling' from one pen to another so that the sheep have no choice but to follow the direction intended. Most important, this flow should not be impeded by sharp corners.

(3) There is considerable virtue in having a circular forcing pen at the entrance both to the dip and to the race. Sheep are not exactly stupid and they have long memories, particularly of unpleasant experiences, and given the choice they will generally prefer not to enter either of these. Again, keeping them up tight makes life a lot easier.

(4) A good race really is essential. Narrow enough to

22. Home-made sheep trailer at Worlaby being loaded. Essential for the easy movement of ewes and lambs.

Source: David Lee.

23. New set of sheep-handling pens designed by Mr. Geoffrey Hyde at Lulworth Castle Farms, showing race and swim-through dip.

Source: Author.

24. Poldenvale handling pens erected for sheep dipping.
Source: Poldenvale.

25. Isa Lloyd 'walk in' dip bath.
Source: Poldenvale.

prevent the average ewe turning round and long enough so that the foot-rot bath, when placed in the race, gives each sheep plenty of time to get its feet well and truly soaked. The race should end in a three-way sorting gate so as to facilitate separation into different lots. Alongside the race, you will need a corridor in which the shepherd can stand, and a shelf on which to put syringes, bottles and so on whilst he is working. This is a small but highly convenient aid to good working conditions.

(5) The dip. The actual size and design of the bath will depend on the size of flock. Remember that the law requires you to keep sheep in the dip for a minimum of one minute in order to control sheep scab. A minute is a surprisingly long time, and most sheep will be in and out in 30 seconds or less given the opportunity. It follows that there has to be a gate keeping them in the bath or, alternatively, it has to be of the long 'swim-through' type, long enough to ensure the necessary immersion time. What is important is to remember that being thrown into a dip is an unpleasant experience. Everything should be done to minimise the risk and the stress by having rounded soft corners, whilst at the same time ensuring that the end result is effective.

SOME DETAILS OF DESIGN

The Foot-rot bath

The essential point is that the sheep must have no alternative than to get their feet really soaked. Obviously, the bath must be long enough. Also, there must be no ledge at the sides, where the bath butts up against the race walls, as this would enable the sheep to straddle the liquid. Then, in the bath itself, the floor should be corrugated along its length so that the ridges of the corrugations force open the two parts of the sheep's foot, thus allowing the liquid to penetrate right up in between to the fleshy and susceptible parts of the foot.

To take our experience at Worlaby once more, we built this in especially hardened *in-situ* concrete. Today you would not do anything of the sort, as it is far more practical to buy one of the many makes of prefabricated baths that are on the market.

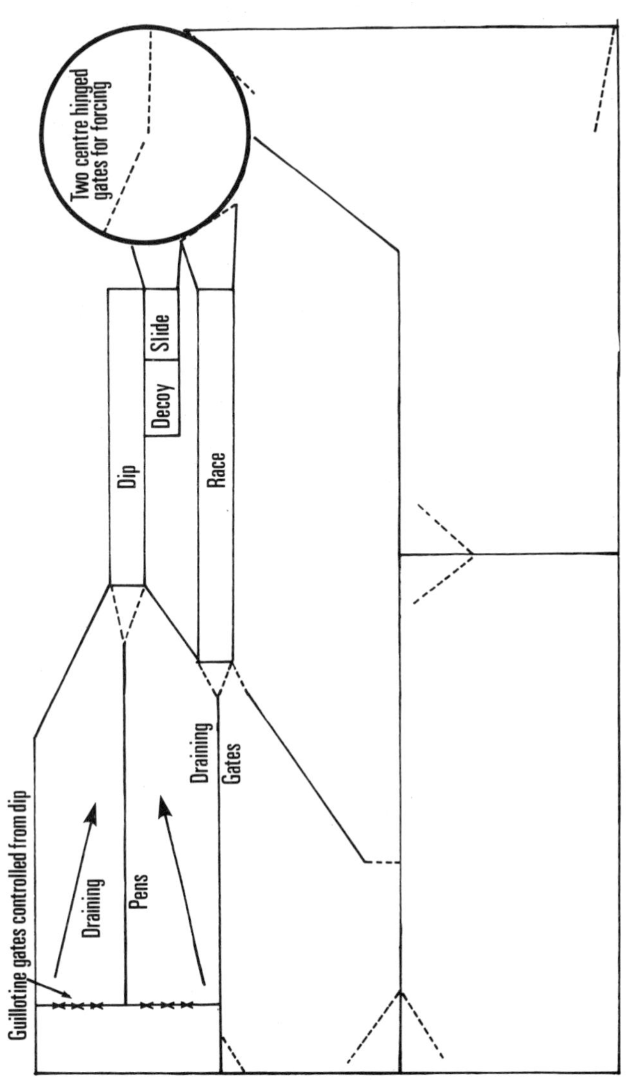

Fig. 5. Layout of Worlaby pens.

These are made either of galvanised metal or fibreglass. I have found the metal ones much more satisfactory and hard wearing. Dimensions vary from one maker to another but all follow the basic principles I have set out. When designing your pens, it follows that the length and width of the race will be determined by the dimensions of the foot-rot bath that you buy. My advice is – go for as great a length as you can, even to the extent of having two baths, the one following the other.

A final point about foot-rot control, which is very much a matter of design of equipment. That is, the job must be easy to carry out. If it is too laborious to set the equipment up; if you have to move the sheep too far; if it is too much of a temptation to say 'we cannot fit that job in today, it will have to wait until we have more time!' – if it is any of these things, then something is fundamentally wrong with the design of your handling equipment. For the essence of foot-rot control is regular treatment. This is a powerful argument in favour of mobile pens. On many occasions during the grazing season, it is much more practical to run the sheep through the bath in the field.

The Dip

I have already made the point that the dip design must be such as to minimise the discomfort and possible injury to the sheep. *No sharp corners.* Having stated that, there are two clear choices of design – for the really large flock, and for the small.

For the *large flock,* where 2,000-plus sheep may be involved, there is a great deal to be said for the 'swim-through' design. What you have to achieve is a combination of (a) high output per hour, (b) making the job reasonably easy for the shepherds, and (c) the statutory immersion period of one minute. The basis of the design is a long narrow trench, narrow enough so that the sheep cannot turn around, down which the sheep swim. At the exit there has to be a ramp up which they can walk, and it is essential that the surface of this is 'ribbed' with cross slates of some kind so as to avoid slipping.

Getting the sheep into the dip demands rather greater ingenuity. I favour the 'side slide in' principle which is part of our design at Worlaby. Coming from the circular forcing pen, the sheep see two ewes in a decoy pen at the further end of a short passage. It is important that they do not see the dip itself. The entrance to the dip is covered with a wooden floor which is side-hinged and counter-balanced by weights. When four average sized ewes are on this false floor, it tips under their weight and slides them gently sideways into the first part of the bath. This entry into the bath is twice the width of the swim trench and is what you might call the traffic control point. One man needs to be stationed here with a dipping crook, making certain that each sheep really goes under. With this arrangement, two men can dip a really large number of sheep a day – well into thousands – without too much fatigue. But the number will depend very much on how long it takes to get the sheep to the pens. Distance again.

Coming out of the dip bath, the sheep will be carrying in their fleeces a large quantity of the dip material. Dip is expensive and you just cannot afford to waste it. So it is essential that there are draining pens in which the sheep can stand for a minimum of five minutes. The floor obviously must be of concrete and must slope back to the dip bath itself. This floor will however get soiled with droppings, so it is important that on its way back to the bath, the liquid should pass through a simple filter.

These, then, are the essentials of an arrangement very suitable for the large flock where ease of operation and speed of throughput is important. It is expensive to construct and cannot be bought in prefabricated form. It also needs a large quantity of dip to fill it, probably 8,000 litres. There cannot be any half measures; you cannot half fill it if, for some reason, you have only got a few sheep to dip. For the depth has to be there in the swim-through trench for it to work at all.

For the small flock, all the same principles apply. Speed of operation is much less important, but consideration for the man or men doing the job remains – you will probably be doing the job yourself! Man-handling a heavy old ewe into a dip bath on a hot day when she is determined not to go in, is not everyone's idea of recreation! One thing you cannot do with a small dip is to so arrange it that you have a continuous flow of sheep through the bath. There has to be an element of stop and start so that each ewe has its requisite time of immersion. Generally, this will involve having some form of swinging, top-hinged door at the exit to the bath so as to prevent the sheep leaving up the ramp.

This type of dip can be bought readily from several different makers in prefabricated form *all of a piece* and usually made of some form of fibreglass construction. The whole thing is then fitted into an appropriately prepared hole. Obviously, there are many different designs, sizes and combinations of bright ideas. Before making any choice, my advice would be to visit as many practical working installations as possible. Go to the National Agricultural Centre's sheep unit at Stoneleigh; consult ADAS and also the leading manufacturers of dips, some of whom issue very comprehensive leaflets on dip design and operation.

OTHER EQUIPMENT

First and foremost, a weighing crate. No one, I repeat no one, who is producing lambs for sale for slaughter should be in business without a means of weighing. I defy anybody, however skilled, to guess the weight of live lambs to the degree of accuracy needed for present-day marketing. There are plenty on the market specifically designed for the job.

Secondly, you might care to consider what is generally called a cradle. That is to say a gadget into which you put a ewe, secure her, and can then turn her upside down by means of a pivot. The object, of course, is to make paring of the feet much easier and less back-aching. In no way are cradles essential, and many shepherds, including my own, won't use them. They prefer to up-end the sheep in the normal way, holding them between their legs whilst doing their feet. Some people, on the other hand, swear by cradles and would not be without them.

Thirdly, a trailer of some sort, useful all the year round, but particularly so when turning ewes and lambs out from the lambing pens to distant fields. I know that you can walk them out and no doubt this is feasible when the grass fields are adjoining the wintering shed or lambing pens, but my word, it is a time-consuming job! I much prefer to have a suitable trailer (see photograph 22) with a ramp and a separate compartment for the young lambs. It need not be an elaborate or expensive affair and can usually be made up from an old trailer body by any handy blacksmith or mechanic.

SHEARING

The ultimate, without any doubt, is what is called the New Zealand shearing shed – a special building so designed as to make the whole job of shearing quick and efficient. Conditions in New Zealand are, however, quite different. They deal with enormous numbers of sheep, and wool for them is a product of relatively far greater importance than it is for us in Britain. I think that it would be very difficult indeed to justify such expenditure here, and in any case there are other, and just as efficient ways, of achieving the same objective.

There are still many people who shear out of doors. All that is necessary then is a set of holding pens, the shearing floor – which only needs to be a large enough area of wooden boarding – and the appropriate stands for the shearing machines, which will be powered either by mobile generator or by petrol engine.

That is simplicity itself. To my mind, it is well worthwhile going to greater effort, not only to make the job easier, but also to produce wool of better quality. Wool must be clean, both from muck and from soil, and it must be dry. Fall down on this, and you will quite properly be penalised by the Wool Board. All this is much more easily achieved in a building, and ideally in your inwintering shed if you have one.

At Worlaby, we use the slatted floor sections that are used around the water troughs (see Chapter 7). They are put together on a base of railway sleepers to form the holding pens prior to shearing. The pen divisions themselves are mobile metal hurdles. Each pen is small, holding no more than 20 or so ewes, and they are immediately behind the shearing board so that the shearer can quickly get hold of his next sheep. In front of the shearers is a large table onto which the fleece is thrown to be wrapped and tied, prior to putting into the wool sheet. All very simple, and made up out of materials already in use at other times of the year. It is well worth while, and equally easy to set up in any building such as a Dutch barn.

'LIVE EQUIPMENT'

This chapter is all about sheep handling: making all the necessary day-by-day tasks easier both for shepherd and for sheep. Easier is almost without exception more efficient. The chapter would not be complete without a mention of the most useful piece of equipment of the lot – the sheep dog. Whether on the open hill where the shepherd just could not cover the ground without his dogs; or, at the other extreme, on a lowland farm like mine, where sheep have to be taken past unfenced arable crops – a well-trained dog can only be described as a godsend. Not only that, but one of the real and satisfying pleasures of life. But it *must* be a well-trained dog – for a bad one is a nightmare!

I would not advise anyone to buy a fully trained dog. For one thing, it would be extremely expensive, and quite rightly so, for it will have taken great skill to get it trained to a high pitch. More important, you and the dog will not know each other – and this is important, for the operation of a working dog, be it Border Collie or Labrador, is a highly personal one. Both are intelligent and very affectionate, and they need a warm relationship with their master to give of their best. Watch the top handlers at a sheep dog trial and you will see what I mean.

So I would suggest that anyone should start with a puppy and train them themselves. That training need not be to trial standards, and indeed for the ordinary lowland sheep farm, that is not what is required at all. It does need to be obedient to the commands of fetching and driving and holding the sheep; and, above all, have a knowledge of the farm and its oddities. There are several very useful books published on the training and working of a Border Collie, and if you are in any doubt at all, I would suggest that you get one of them.

Chapter 13

THE SHEEP CALENDAR

I HESITATED for a long time before I finally decided to write this chapter. I have such a dislike for anything that might be considered a livestock blueprint, something which is so open to misuse and misinterpretation, that for some time I felt it was inadvisable to attempt a calendar. As I have said elsewhere, success with livestock depends on observation and anticipation as well as a following of the rules – the animal equivalent of green fingers. However, I eventually decided to write it because of the importance of a number of key points in the sheep year. These are times when it is critical that things go right and when the effect on the eventual harvest is considerable. So this is not a day-by-day, or even a month-by-month, account of the year's work. Rather it is a discussion of those key areas and why they are important.

PREPARATION OF EWES FOR TUPPING

It seems right to start the year at this point in midsummer. Weaning of the lambs is a natural break, not the end of one harvest but certainly the beginnings of the preparation for the next. The traditional practice used to be to turn the ewes away onto poorer grazing, at high stocking rates – to put them into a sort of 'concentration camp'. The thinking behind this was partly correct, but mostly wrong, if I may put it like that. Obviously, ewes which are 'on holiday', after having finished feeding their lambs, and before they go back to the ram, have considerably reduced needs for food. They have only got themselves to look after. So far so good, but the second part of this reasoning was that in order to 'flush' the ewes for tupping, they had to be brought right down in condition so that they could be pushed up again. The theory is that the rapid rise in

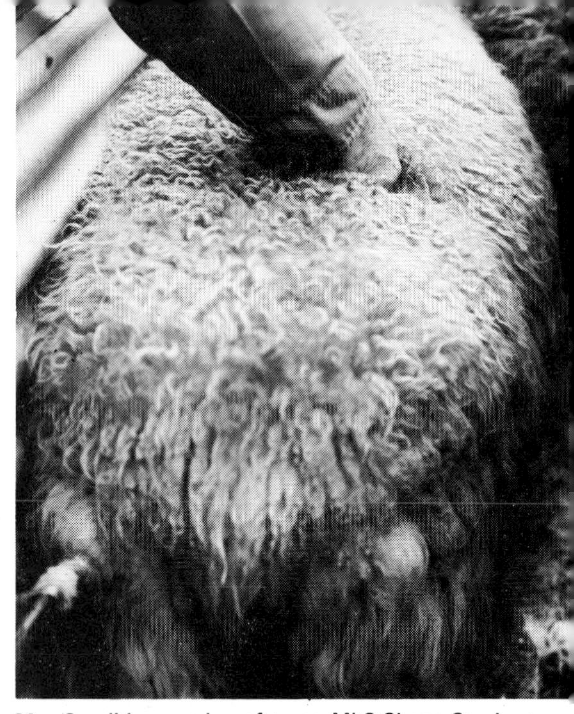

26. Condition scoring of ewes. MLC Sheep Services.
Source: Farmers Weekly.

27. Selection of lamb for slaughter. Brynmor Morgan, MLC sheep specialist for Wales, advising on handling to assess carcase quality.
Source: MLC.

28. Vaccination. The correct site for vaccination so as to avoid the risk of abscess formation in the valuable cuts of the carcase.

Source: MLC.

29. Paring of sheep's foot. Essential regular treatment avoids the build-up of infection.

Source: Farming Press.

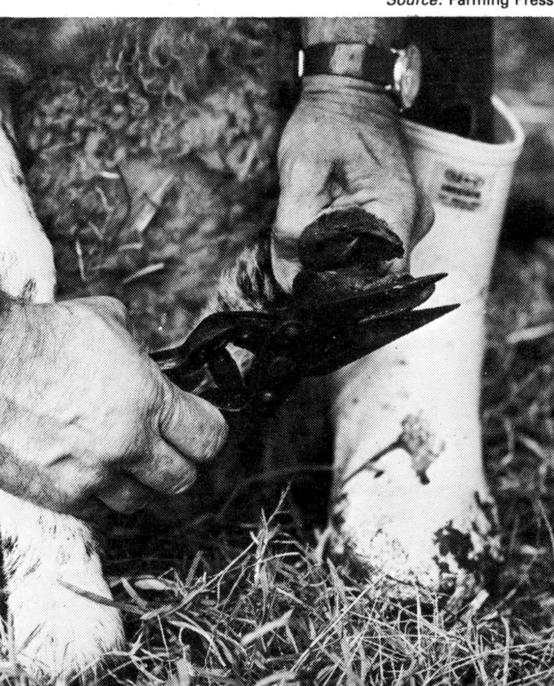

condition from relatively poor to good, which is called flushing, leads to the maximum number of eggs being shed at each ovulation, thus leading to a high lambing percentage. In my view, this is open to some doubt at the very least, and in any case is far too simple an approach.

Certainly the ewes must be in 'good' condition at tupping time. Ewes which are poor for reasons of either sub-standard feeding or bad health will protect themselves automatically, either by not conceiving at all or by only conceiving one lamb. That goes almost without saying. Equally, I believe that ewes must not be mud fat. If the female's internal organs are covered with fat, then she is not going to function properly. This is straightforward commonsense which is as true for sheep as it is for cattle, or human beings for that matter.

So what is 'good condition' and how can we achieve it? There is a method of describing condition now known as 'condition scoring' which has been developed by MLC (see MLC leaflet, *Body Condition Scoring of Ewes*). This is quite simply a scale of 1 to 5; '1' being extremely lean and '5' equally extremely fat. These are descriptions of the 'feel' of the ewe's back at each stage, which enable anyone to locate a ewe in the appropriate place on the scale.

Now there is no magic about condition scoring. It solves no problems – it merely tells you what you have got. There is a temptation to condemn it as a fancy piece of theory. I have often heard experienced shepherds or flockmasters pooh-pooh it in scathing terms, saying that anyone who is worth his salt would know what condition his ewes were in without putting it on a scale. 'What nonsense from these boys straight out of college', etc! This does less than justice to the facts of the matter. Condition scoring is in reality a discipline. In order to score the ewes, you have actually got to handle them *all* and put them into different categories. It is this single fact which is the virtue of the exercise.

Let us go back to the point of weaning. Some ewes, hopefully the majority, will have been feeding twins. Some will be much better milkers than others. Some will have had some health problem, serious or slight. However good the management of either sheep or grass, the complete flock will consist of individuals whose condition range from the very good to the very poor. It should be one of the urgent jobs to do, as soon as the ewes have lost their milk and settled down after weaning, to go through them and physically separate them into, probably, three different flocks: good, average and poor condition. The objective should be to arrive at the pre-ordained date for putting the tups in with *all* the ewes in good condition. Let us say, at condition score 2½ to 3. Given that, at weaning, the ewes will be in all sorts of different condition depending upon their performance, it is quite obviously impossible to get them all into the same state without separating them and treating them differently. The poor ones must be given better grazing; the best ones can afford to be put on a more severe ration. Each according to her state and regulated accordingly.

In my experience, this is an essential part in achieving a high flock output. High prolificacy under practical conditions is not achieved by having lots of triplets; it is achieved by having very few barreners and by nearly all the ewes having twins. In other words, having every individual performing well is far more important than having some of them performing at extraordinary levels. This work starts immediately after weaning when the foundations of next year's crop are properly laid.

Has the concept of flushing in modified form got a part to play in this approach? Certainly not in the form of taking the ewes down in condition. Nor is there any justification for thinking that a rapidly rising plane of nutrition in the last three weeks will materially affect the lambing percentage. Only if for some reason, such as a severe late summer drought leading to shortage of food, will it be worthwhile providing extra nutrition for this final 'push'. Under normal conditions I believe that we can forget flushing and concentrate all our energies on getting every individual right and on a steady plane of nutrition.

THE MANAGEMENT OF THE RAM

Why is it that otherwise excellent shepherds will neglect their stock rams? It is quite extraordinary how often one sees this. After the job is done, they are pushed away into some old paddock and left to their own devices. Usually it is old pasture and has been used for the same job year after year, so they get full of worms. Feeding is often second-rate; and foot-rot control is something which is forgotten about until the last moment. Am I exaggerating? I don't think so, but if none of this applies to you, my apologies!

Whatever the case, it really does not make much sense to neglect this so-important half of the flock just because they only work for six weeks a year. Good nutrition and good health, particularly good sound feet, are surely obvious requirements. But there are some matters of detail which merit further consideration.

Obviously, a ram is worth nothing if he is not fertile. It is surprising how common it is to find one that for one reason or another has infertile semen. Generally, this is not noticed because most people run rams in groups of at least two or three, and the others will cover for one bad one. To some extent this does not matter, but it does depend upon how many ewes have been allocated per ram. I think it is worthwhile making a check. Semen testing is now much easier than it used to be and many vets are equipped to do the job. Infertility may be congenital, and there will be nothing else you can do but to scrap the ram. There are, however, some causes of purely temporary infertility which are easily enough avoided but rarely seem to be.

For example, use of most antibiotics, for whatever reason, will almost always make a ram infertile for six weeks or so after treatment. Few people in my experience seem to know about this. Then, there is the case of certain types of tick-borne fever which will affect a ram newly introduced onto the farm and which has not had the time to gain immunity. The remedy is obvious – have the rams on the farm well before they are wanted.

This leads me to say something about dates of purchase. I have been producing shearling rams for sale for a good number of years now, and I always have the greatest difficulty in persuading customers to take delivery of their rams soon enough. It is, I suspect, the result of some meanness on their part, preferring that I should continue to feed the rams as long as possible. If it is, then it is surely false economy. Any animal takes time to settle down to a new environment. The journey itself will impose a degree of stress. The new home will have a different climate, different grasses, different mineral levels, and certainly different bugs. All this can lead to temporary infertility – so please take your rams at least two months before you need to work them, in your own interest.

Infertility I have taken so far to be infertile semen. There are cases where a ram will not serve ewes on heat. This can, of course, be due quite simply to sore feet, particularly hind feet. Sometimes, however, young rams will not serve, either due quite simply to inexperience or because they have become somewhat homosexual in their shearling year, having been kept in a group of rams of the same age. This is sometimes somewhat irreverently referred to as the 'Public School Syndrome'! Be that as it may, the cure is generally quite simple. Shut the gay young ram up with an old ewe who is on heat. It never fails!

Incidentally, it ought to be a matter of concern to all ram breeders that they should try to avoid this state of affairs. There is some evidence to suggest that homosexuality is much more likely to arise when young rams are kept together in large numbers. Thirty to 40 together in one bunch is probably the limit.

Lastly, there is the all-important question: how many rams should I have? The generally accepted figure is between 40 to 50 ewes per ram, and down to perhaps 35 for ram lambs. This was certainly the figure which I used to use in the days when I was buying rams. Recent experience has suggested to me that we are putting ourselves at a disadvantage by asking too much of our rams. As a seller of shearling rams, I obviously have large numbers of ram lambs on the farm. There is no reason why they should hang about doing nothing. Indeed, from the point of view of giving them experience, thus avoiding the problems of developing homosexuality as well as of proving them, there is much to be said for using them. So we use our whole ram lamb flock on our commercial ewes, which works out at around one ram lamb to every 15 ewes. Now I am not suggesting for one moment that any commercial lamb producer should go to quite such lengths as that. But I can say that using such a high proportion of rams has had two significant advantages: hardly any barren ewes and a lambing which is 85 per cent completed in two weeks. This was a long way from being our experience when we had one ram to 50 ewes and it does suggest to me that one ram to 30 ewes might be about the sensible average.

THE IN-LAMB EWE

In another section of this book, I have stressed the importance of two separate periods during the ewe's pregnancy: the first eight and the last six weeks. I make no apology for returning to the subject.

The first eight weeks, during which time the foetuses become settled in the uterus, are critical. Any nutritional stress on the ewe during this time and she will automatically react defensively so as to preserve her own well-being. At worst she will abort. Under less serious stress, she will absorb a foetus and reduce her load to a single. Let me emphasise that the stress does not have to be serious for this to happen. With most commercial flocks, that is to say, those flocks that are lambing in March, this period will coincide with the onset of bad weather and rapidly deteriorating quality of grass. A period of 10 days or so of cold, driving rain is all too common in late November, and my guess is that it is quite enough to drop prolificacy by a very significant amount.

Then, the last six weeks. The number of lambs to be born is long since fixed. What is not fixed at all is how viable they are going to be at birth; how you feed the ewe during this period will determine that. I have already dealt with the detail of this fully in Chapter 9. Suffice it to say here that losses of live lambs during the first 48 hours of life can be staggeringly high. Almost all these losses are due to either or both, lambs being born weak and the ewe having little or no milk. By the time that is happening, it is too late to do anything about it. Don't leave it that late!

LAMBING

Whether your ewes have been in or outwintered, there is general agreement that it is worth going to a great deal of trouble to provide suitable penning arrangements for the actual lambing time. This may take the form of a lambing yard constructed in a field, probably a rectangular enclosure made of straw bales and sheep netting and surrounded by individual pens of bales roofed with corrugated sheets. Or it may be a more permanent arrangement in buildings where, if the sheep are outwintered, they are at least brought in for lambing. Or it may be in the form of lambing pens built in the wintering sheds.

Whatever the case, the principal objectives remain the same. That is to say that, immediately after she has lambed, the ewe should be placed in an individual pen with her lambs. She should remain there, I think, for a minimum of 48 hours so that the mother and child bond may be firmly established. At the same time, the routine operation of navel dressing with an antiseptic spray, tailing and castrating will take place. Be sure that she has ample food and water and don't forget the need for high protein levels (see Chapter 9).

How many pens do you need to have? My rule of thumb is that there should be sufficient to be able to keep ewes and lambs penned up, at peak lambing pressure, for a maximum period of five days. Really bad weather, particularly in March, is not likely to last longer than that. The ewes can then be turned out either directly to grass, or into bulking-up pens.

The point at which the ewe and her lambs are transferred from their individual nursery is one where serious losses occur. Mismothering, whether caused simply by crowd pressure or by the ewe going off over enthusiastically to have a fill of appetising grass which she has not tasted for many a long week – whatever the cause, a lost lamb does not have to be away from Mum for long before it is a dead lamb. So anything that can be done to keep mother and offspring together is worthwhile.

The use of bulking-up pens, holding perhaps no more than 20 ewes at a time so that everyone gets used to living together; turning out to grass in small numbers into different parts and corners of different fields; and regular visiting to 'mother up' separated families. All these points are well worth implementing. This last and most important chore is much helped if ewe and her lambs have the same number spray paint marked on their sides. Remember that the last few lambs, like the last kg of wheat per hectare, are really profitable ones. Losses hurt. Perhaps the biggest danger of all is that, under the pressure and fatigue of lambing, shepherds can become inured to losses and accept them as inevitable. They are not.

Talking of lamb losses, particularly those occurring during the first few hours of life, there is one gadget which is capable of saving more lives than any I know. This based upon the certain fact that unless a new born lamb gets colostrum inside it very quickly indeed, its chances of survival are much reduced. A survey of lamb losses which occurred during the first 24 hours of life, carried out by the MLC showed that virtually all of them had completely empty stomachs. The gadget is an ordinary human catheter, which you should be able to persuade your doctor to give you for a few pence. Attach a 50 cc syringe and you have the means whereby you can introduce colostrum into the lamb's stomach.

The procedure is as follows: First get your colostrum; that from a ewe is, of course, the best. In many cases you will be able to pinch some from a heavy-milking ewe just lambed, to help out a less fortunate neighbour. But to be on the safe side, we always beg some cow's colostrum from a nearby dairy farmer and put it in the deep freeze in small plastic bags. There it will keep indefinitely and can be brought out as required.

Which lambs should you force-feed? We do all triplets as a matter of course, and any doubles that look at all weakly. And, of course, all lambs whose mothers lamb down with no milk. What you do is this: introduce the catheter tube slowly and carefully into the back of the lamb's throat, allowing it to swallow the tube. Go on feeding the tube into the lamb until 30 centimetres of tube have disappeared. It sounds difficult, even horrific, but it is surprisingly easy once you have the knack – get someone who knows how, to show you. Then attach the syringe, suitably filled with warm colostrum, and gently syringe it into the lamb's stomach via the tube. You will probably only have to do it once, perhaps a second time after three or four hours with weakly lambs. I can promise you that the results are miraculous.

What about 'spare' lambs, whether they be orphans or one of three that you do not wish to leave with the ewe? By far and away the best thing to do is to get ewes that have only had singles to adopt them. This is easily done, particularly if a triplet birth coincides with a single, when the spare lamb can be transferred almost without the new mother noticing. If possible, rub the lamb to be adopted in the afterbirth liquid, so that it smells right. The more difficult cases occur when you have a ewe who has lost her lambs and who is well past the stage where she can be fooled into accepting other people's kids. There is the old shepherd's trick of skinning her dead lamb and putting this skin as a jacket over the lamb to be adopted. If this does not work – and it often does – then you will have to resort to an adoption cage. A device where the ewe is secured by the head so that she cannot butt the lamb, but allowing it to suckle. After a few days, even the most ill tempered of ewes will generally give in and accept the lamb.

Lambs can be reared artificially relatively easily. There are now several brands of ewe milk replacer on the market which work well, and where the reconstituted milk can be fed cold. The rules for success are well set out in MLC booklet, *Artificial Rearing of Lambs.* I suggest you study it well, for if you are going to do the job, then you must do it properly.

Hand rearing, half done, is a waste of time and money. In my experience, women are much better at it than men, and it is an ideal job for the shepherd's wife if she is so inclined. A word of warning about the economics. It is not done cheaply and if you have to pay for labour as well as for the food, then the present relationship between slaughter lamb prices and food costs leaves a very narrow margin. If you are in an area where spare lambs can be sold, and when, as in 1985, these could be sold for between £12 to £14 each at around a week old – then my advice would be to take the money and let someone else bear the cost.

OUT AT GRASS

I have already dealt with the arguments on set stocking versus the various forms of creep feeding. This is the point to re-emphasise the desirability of a quiet life. Various health treatments will no doubt be necessary during the time whilst ewes and lambs are together at grass. Collection and penning will be inevitable, and the resultant stress will be certainly a set-back to growth. But do keep it to a minimum. In particular, restrain both yourself and your shepherd from using a sheepdog. The dog will be there, will be full of energy and longing for a run, and the temptation is strong to let him go. Resist the temptation. A well-trained dog is there to help you catch an individual sheep, should that be necessary.

A question about feeding at grass. Should the lambs receive some form of supplementary concentrate food? Obviously, they will be getting ample protein both from milk and from the grass, and a case could only be made for feeding some cereals as a supplement. It is, of course, all a question of economics. Will it pay? The answer to that will depend very much on the system you are following. If it is to lamb relatively early and you are seeking to sell the bulk of your lambs early in the season before the market price drops, then it almost certainly is justified. If, on the other hand, you are a late lamber and a late seller, your interest will be to keep costs down to the minimum. Creep feeding certainly puts a bloom and a fatness onto lambs which is attractive to see – but it is worth nothing until it is turned into cash.

WEANING

And so we come full circle. Remember that July 1st is the critical date. Things can go quickly and badly wrong with lambs if they are left with the ewes after that; certainly at high stocking rates. The ewes' milk yield will have fallen to low levels, the lambs will be dependent upon grass for the bulk of their food, and hence will be in competition with the ewes for the available supplies; grass quality will be falling away rapidly, no matter how good your grassland management. The ewes will be churning out worm eggs onto the pasture.

All in all, a powerful incentive to make sure that you have adequate food available onto which you can put weaned lambs.

Chapter 14

THE CONTROL OF DISEASE

DISEASE IN all its various forms, its prevention and control, is part and parcel of the daily work and planning of any stockman, whatever the species. Whilst with sheep we are not pushing the intensification of management to unhealthy and artificial limits, it is nevertheless true, that as we go up the scale towards 15 ewes to the hectare, as we push up lambing percentages towards the 200, and as we intensify grassland production, we need to be more conscious of disease prevention. To this extent I believe that it is absolutely essential that the modern stockman should be well versed in the knowledge of disease, its prevention and cure.

I would be the last person to say that the vet has not got a major role to play in the management of a flock of sheep. Quite the contrary, I wish that more vets were willing and able to involve themselves in routine preventative visiting so that problems might be anticipated. Too often the vet is called in to act as a fire fighter, sometimes when the blaze has got out of control. The fact that more vets do not participate in management is not always the fault of the flock owner. I have been fortunate in that I have always had an extremely good service, both from our own local vet and from the Veterinary Investigation Centre, nearby at Lincoln. But it has to be said that there are too many vets around who have not been able to keep abreast of the developments in sheep management, and whose outlook is attuned to a much less intensive age.

So I insist that you should call the vet too soon rather than too late; and preferably that he should be knowledgeable about the day-to-day management and in tune with it. But he cannot be there the whole time and it would cost you a fortune, quite rightly, if he was. There is much that the shepherd/flock-master must know and do for himself. Just as he should know almost instinctively when it is beyond him and he needs to call in professional help. This chapter is written in that spirit. It does not pretend to be a do-it-yourself animal health manual. That would take far too much space. The Department of Agriculture and Fisheries for Scotland used to produce a small booklet entitled *The Shepherd's Guide.*

Another publication which is certainly available comes from Liverpool University. With the somewhat daunting title of *Notes for the Sheep Clinician,* it gives a detailed cover of all sheep diseases, their symptoms and the treatment. It is not, and perhaps could not be, written in simple language, but it is well worth having, nonetheless.

EVERYDAY DISEASES

So I want to devote the bulk of this chapter to a resumé of those diseases, disorders, or deficiencies which are almost inevitably the lot of every flockmaster; those which he should seek to prevent whenever possible by his management and stockmanship, and which he should be able to control and cure almost as a matter of routine. I want to underline the importance of stockmanship. Disease always strikes worst when animals are in a low condition. I do not say necessarily a poor condition, because sheep, like us, can become unhealthily fat, but a low condition which is a state which falls short of thriving and is often associated with stress.

(1) *Worm Parasites*

There can be no doubt that this is where we must start, for there is no other disease which causes so much trouble and which is so intrinsically part and parcel of intensive sheep management. So much so that the oft-repeated saying: 'The worst enemy of one sheep is another sheep' was used as an argument against intensification. We cannot accept that. It is not necessary that we should, nor indeed could we afford to.

Fortunately, there is an excellent range of drugs on the market and their regular use has to be part of our armoury of control. But, first and foremost, we must look to management to reduce the incidence of infestation. Although actual disease and death are important, of course, as sources of loss – the symptoms are obvious and lead to immediate treatment – probably far greater losses result from those lambs showing no symptoms, but which, nevertheless, are carrying a moderate worm burden with resultant slow maturity and failure to thrive. We really cannot afford to accept this. Management control of worm infestation is based upon the following points of attack:

(a) Any condition which lowers the resistance of the lamb favours the worm, and a heavy worm burden can build up as a result. The primary cause may therefore be found amongst such things as cobalt deficiency, a very general mineral imbalance, overstocking, or quite simply a sheer shortage of food.

(b) Pastures can become heavily contaminated with the worm larvae, and whilst many of these will perish under favourable conditions, of sunlight, drought or extreme cold, yet many will survive from one year to another. Putting sheep back on to these pastures merely ensures reinfestation. Hence a pasture becomes 'sheep sick'.

(c) Very few of the worm species which affect sheep have a common host in cattle. Grazing a pasture with cattle can therefore break the cycle and clean up the pasture for the sheep. Even more effective is leaving the field free of both cattle and sheep and using it for conservation for a whole season.

Guidelines

Out of these three facts comes a logical set of guidelines for sheep management which will do much to minimise worm damage.

(1) Never allow the flock, and in particular the lambs which are most susceptible, to be subject to other stress, be it mineral shortage or sheer malnutrition.

(2) If you are in an arable context, then there is a great deal to be said for restricting the life of your leys to two years. It is particularly in the third year of intensive stocking with sheep that you will really start to run into problems.

(3) If your farming system involves long-term leys or permanent pasture, then you should give very serious consideration indeed to the clean grazing techniques developed by the East of Scotland College. A three-year

rotation of sheep, followed by cattle and finally by a year's conservation, works wonderfully well and is really the only practical way of intensifying output from the long-term grass without running into serious worm problems. An essential part of the success of this system is that the species must not be mixed.

(4) Wherever possible, make use of clean crops for lamb feeding, such as turnips, swedes, rape and kale (see chapter on Forage Crops).

(5) Finally, sheep grass should be kept short, not only because it is more nutritious that way, but also so as to allow sunlight and drying winds to penetrate the sward, and thus kill off the maximum of worm larvae.

When, however, you have so managed your sheep and your grass and forage crops so as to keep trouble to a minimum, you will still have to live with the fact that worms of various types will be with you. So the use of anthelmintic drugs is essential. I am not going to give any suggested programme for dosing. The problem must be discussed in the context of the individual farm, its geographical location, its system of management, etc. There are, however, I think, three general points worth making.

Firstly, with inwintered ewes, there is considerable evidence to support the practice of drenching six weeks before lambing. This cleans the ewes out at a time when the worms are in a susceptible stage of their evolution and gets the grazing season subsequently off to a good start.

Secondly, it is well worthwhile ringing the changes between the different types of worm drench that are available. Each has its strengths and weaknesses against different categories of round worms, and one can complement the other. The exception to this is where there is a specific problem, for example with liver fluke or with *Nematodirus* or with *Coccidiosis,* when the specific product must be used.

Finally, there is the rather odd situation which often seems to happen when a few odd sheep, both ewes and lambs, do not seem to respond to a recent drench. The bulk of the flock is doing well, but a very few can be scouring for no apparent reason. I suspect that this is often due to these particular sheep having held the drench in their mouth and then subsequently spat it out. Whatever the reason, they are there showing obvious symptoms. I think the best way to treat them is for the shepherd to catch them individually – which he can easily do with his dog in the field – and treat them with a Thibendazole tablet. These tablets are very convenient as they can be part of the gear which the shepherd carries round with him in his bag.

A final word on worms. Don't go mad with the drenching gun! Anthelmintics are very efficient, but they clean out the sheep's insides at a cost. There is bound to be a check to the lamb growth and of course this is worthwhile to get rid of the greater evil, which is the damage the worms are doing. But don't turn it into a routine which you do regardless of whether it is really necessary or not.

(2) *Skin Parasites*

The real worry is Sheep Scab. This wretched parasite, which if allowed to go unchecked will so emaciate a sheep that it will die, can be controlled and eliminated altogether quite simply by dipping with the approved dip during the statutory fixed period in the autumn. For many years, Britain was free of this pest and we only have it now because we were stupid enough to allow it to slip in in live sheep imported from Ireland. It is now established again, but it could easily be got rid of. Unfortunately, it keeps recurring, and this can only happen because some sheep are not being dipped. The finger of suspicion is pointed very firmly towards the hill areas of Dartmoor and Exmoor, where one can only suppose that some people are so criminally stupid as not to bother to dip some, or all, of their sheep.

I think the main reason why it is nowadays proving to be so difficult to stamp out Sheep Scab is that in the previous campaign it was the village policeman who had to be present at the dipping. It was almost impossible to fool the local 'bobby', who knew very well what was going on in his own parish. Now that that worthy individual no longer exists, and has been replaced by the Panda Car from the nearby town, it is all too easy for some irresponsible individual to break the law. And now that sheep can be moved over long distances by fast, big lorries, a local outbreak can be spread over the whole country in no time at all. We shall only finally knock this scourge on the head when we get back to local supervision of all dipping.

Fly strike is the other main reason for dipping, but, of course, at a totally different time of year. I well remember as a boy going round twice a day with the shepherd with a bottle of Jeyes Fluid, seeking out those sheep that had been struck, catching them and flushing out the maggots with the disinfectant liquid. A nasty job and, in a real sense, shutting the stable door after the horse had bolted. Thank goodness for modern dips which now kill off the young maggots as they hatch in the wool.

There are other parasite conditions which warrant a mention. Sheep head fly can be an infernal nuisance, particularly in Scotland and the North of England, where the particular fly exists. It attacks the head and causes sores which invite further attack. It can cause a great deal of suffering and loss of condition. Dipping is only partly effective as a deterrent. Dressing of the wounds with a lotion that is both fly-repellent and healing is helpful but time consuming.

Then there are ticks which are only a problem where there is growth such as heather and bracken. They can cause serious illness and abortion, resulting from tick-borne fever, and can cause temporary infertility in rams. Sheep can become acclimatised to the tick-born organism, and those farming in tick areas should take local specialist advice.

Then, finally, a mention of a skin/wool condition which appears to be becoming more widespread and more damaging. This is mycotic dermatitis. Sometimes known as wool rot, the organism causes a yellowish scab on the skin which rises with the wool. The subsequent discoloration of the wool, together with a reduction in wool yield, can lead to considerable loss. We have had experience of this and have found that autumn dipping, compulsory anyway for scab, coupled with a dusting in November, down the back of the sheep with Coopers MD Powder to be an effective cure.

(3) *Foot Rot*

Along with parasitic worms, foot rot is unquestionably the most important cause of sheep failing to thrive. Lame sheep are bound to lose condition and thus money. Almost all cases of lameness in sheep are caused by the foot rot organism. A few cases are caused by some physical injury and under wet conditions, lambs particularly can go lame with scald, which is a much less serious condition.

Theoretically, foot rot can be eliminated from the flock, and certainly the savings in time and cost, to say nothing of lost growth, would be very considerable. But in practice, it proves to be very difficult. Routine and regular paring of the feet so that they never become overgrown, and thus an easy harbour for infection. Regular passing through a 10 per cent formalin mixture in the foot rot bath. All these are necessary, and these jobs must be carried out too often rather than too little. We have found that vaccination, twice a year, certainly doesn't eliminate it. And it is expensive. Control of foot rot is a steady grind, and a job that must never be neglected.

(4) *The Clostridial Diseases*

These include lamb dysentry, enterotoxaemia, black disease, black quarter, braxy, tetanus, and pulpy kidney. All caused by different types of Clostridium bacteria. They used to be real killers, and the fact that they are no longer so is due to the use of highly efficient vaccines. Properly used, these vaccines make these diseases almost something in the past. Pulpy kidney can become a cause of loss, particularly when lambs are 'doing' very well on rich food, but it is almost always associated with the sheep running out of 'time' – the cover given by the previous vaccination having been exhausted.

(5) *Abortion*

It is by no means uncommon to have a few ewes slipping their lambs a few weeks before they are due. There can be, and in most cases there are, reasons for this which give no particular cause for alarm. But I know well the sinking feeling that one has when this happens. Is this the beginning of something much more serious and are we in for an abortion storm? For at that stage there is precious little that can be done about it, and the prospect *could* be the loss of a substantial proportion of the lamb crop.

Abortion can be caused by a good number of different infective organisms. Some like enzootic abortion of ewes can be controlled by vaccination. If there is any reason to suspect its presence, or if sheep are bought in from an infected area, and it is common in South-east Scotland, then there should be no question, the ewes should be vaccinated. Toxoplasmosis, on the other hand, is quite different in that there is no vaccine, indeed much too little is known about it. Then there is abortion caused by tick-borne fever, by salmonella and by vibrio abortus. In all these cases, the only symptom is the aborted foetus. Whatever you believe the cause may be, don't delay – get the foetus off to the Veterinary Investigation Centre immediately for a diagnosis.

Abortion can result from purely physical causes. Crushing through narrow gateways or jumping over obstacles, or even troughs, can cause it. The lesson here, and it applies with particular force with inwintered ewes, is that you should pay serious attention to layout and feeding arrangements, so as to minimise these losses. We had a striking illustration of this at Worlaby, which taught me a lesson. Previously we had used hay racks which served for hay alone. When the time came for starting concentrate feeding, we placed ordinary feed troughs on the floor of each inwintering pen. At feeding time, there was the usual excitement, and as the concentrates were placed in troughs there was a great deal of tearing around and jumping over troughs. I always regarded this with some alarm, and eventually we changed our rack design so that it incorporated a concentrate trough. The racks, of course, formed the outside enclosures of the pen, and it was no longer possible to have the same commotion at feeding time. Our incidence of what I call physical abortions dropped very significantly indeed.

(6) *Orf or Contagious Pustular Dermatitis*

This disease, which used to be rare but is now all too common, is caused by a virus, and can spread rapidly. It seems to occur seasonally, worst of all just after lambing and then later in the summer. It causes scabby sores on the lips, and this can be transferred to the ewes' teats, and is sometimes seen on the feet as well. The sores become swollen and obviously painful. It is particularly serious with ewes and young lambs, as it can lead to premature weaning of the lambs and actual starvation. Prevention is by the use of a vaccine, which is a live one and must therefore be used with care. It has a limited life and is usually therefore given only a few weeks before lambing. Correctly used, we have found it to be efficient although not always 100 per cent. I believe that we are almost at the stage where orf vaccination is becoming a routine, if replacement ewes are bought in.

Treatment is laborious in that the affected sheep have to be caught at regular intervals, probably at least every two days at the outset. The scab crust should be sprayed with an aerosol antibiotic.

The real seriousness of this disease is with young lambs. Whilst not being fatal in itself, losses due to starvation can be high. And the time involved in treatment is very considerable. Once you have got it, it is difficult to eliminate, as infection seems to remain on the farm, no doubt lodged in the wood of food troughs and racks.

(7) *Metabolic Disorders*

There are three which any experienced shepherd will be well aware of. It would be rare indeed if he had not come across them.

(i) *Pregnancy Toxaemia (Twin Lamb Disease)*

This affects ewes in the last few weeks of pregnancy and they are almost inevitably those carrying more than one lamb, hence its name. The affected ewe goes down, often into a semi-coma, and death will occur quite quickly in the absence of treatment. Very often several cases occur together, almost as if one were faced with an outbreak of an infectious disease.

Treatment is really not very satisfactory. Injection of glucose into a vein, feeding a glucose solution, good care and nursing all play a part. Often sacrificing the lambs via a caesarian section is a satisfactory cure for the ewe, but scarcely a productive outcome for the farmer.

Prevention is much more constructive. Twin lamb disease is without any doubt a fault of management – although for practical reasons, it may not be possible to avoid the fault. It is connected with the peculiarly difficult nutritional requirements of the ewe in late pregnancy. With a digestive capacity which is rapidly declining due to increasing pressure from the heavily laden uterus, the ewe eats less at a time when she needs to be eating more. It is of the greatest importance that, during these last six week before lambing, the food she is given is low in fibre, high in digestibility. It follows that roughage should be of high quality and even then should be strictly rationed, so as to allow an adequate consumption of concentrates.

All this is possible with inwintered ewes and really it can be said that pregnancy toxaemia is totally avoidable under these conditions. With outwintered ewes, the position is not so simple. Feeding can be controlled in the same way but the exposure to weather can alter the amount of 'net' nutrition available. Cold driving rain, particularly, makes heavy demands upon the ewe's constitution, and much of the energy which should be used for the benefit of her developing foetus is used up in simply keeping her alive and warm. This fluctuating level of 'net' nutrition which occurs as a result of changeable weather is a primary 'trigger' for twin lamb disease.

(ii) *Lambing Sickness (Milk Fever)*

Closely allied to the same condition in the dairy cow, it has similar symptoms. An acute and rapid fall in the level of blood calcium causes spasm and coma. Treatment is spectacularly successful in most cases, provided they are caught soon enough. An injection of calcium borogluconate under the skin works wonders.

(iii) *Hypomagnesaemia (Grass Staggers)*

In this case, the immediate cause is an acute shortage of magnesium in the blood. The symptoms may be acute, in which case a ewe apparently healthy an hour before, is found dead. Or they may be chronic, when she may remain in a coma for a day or more before dying. Cases usually occur after lambing and at any time up to about the end of May. Some farms, and especially some land types (chalk soil, for example), seem to have a much higher incidence. Weather, too, plays a part as an onset of cold and wet weather after a period of fast growing conditions seems to act as a trigger. But hypomagnesaemia is definitely a condition associated with intensification, particularly with highly productive, lush young pastures. And its increase is more severe with upland or hill ewes when they are put on such pasture than with lowland ewes. There is therefore an element of stress which can play a part.

There is not too much that can be done to prevent the disease occurring by way of altering the basic cause. Certainly there can be no question of deliberately producing poor pastures just because high quality grass is usually associated with low blood magnesium. But there is one element of grassland management which is important: there is clear evidence that high levels of potash recently applied can be part of the trouble. A good rule is always to apply the necessary potash in the autumn, never in the spring.

There are, however, preventive measures which can be taken. The cause is shortage of magnesium and if this element can be got into the sheep in adequate quantities, then all except the most severe cases can be avoided. The problem is how? Magnesium is very unpalatable, and sheep and cattle will avoid it unless it is well disguised. Calcined magnesite at 28 g per ewe/day can be fed in the concentrate ration, but on good grassland there should be no need whatsoever to be feeding the ewes by hand. Just to make it available as part of the mineral supplement in a free-access mineral box feeder is not satisfactory.

There are, however, two other methods, both of which work well. The first is marketed by several of those firms who sell liquid urea feeds. Using the same ball feeders, they provide a molasses/magnesium mixture which, due to the sugar, is palatable and readily taken by the ewes, particularly if they are already accustomed to using this type of feeder. The other alternative – and without doubt the most efficient method where the disease is likely to be severe – is to use magnesium bullets. These are sold by Chapman &

Frearson of Grimsby, following research and development work by the Veterinary Department of Glasgow University. It is marketed under the name of 'Rumbul'. It sounds a bit gimmicky, but it works. The principle is that the bullet, which is heavy, lodges in the animal's stomach and slowly dissolves over a period of five to six weeks, thus releasing a daily quantity of magnesium into the blood system. It is fairly expensive but the cost is nothing compared to the value of ewes which may be lost, and it is something which I would strongly advise on those farms where grass staggers is a near certainty.

Is there a cure if you find ewes going down? The answer is yes, and in the same form as with milk fever, and in this case it is a solution of both calcium and magnesium which is used. Quicker results are obtained if the solution is administered intravenously. Even so, the recovery is generally much less dramatic than with milk fever. The real trouble is that you may never have the opportunity.

(8) *Problems at Lambing Time*

Quite obviously, if there is any time of the sheep year when strict attention to hygiene is important, it must be at lambing time. Both the ewe and her lambs are particularly vulnerable to infection, and real trouble can build up if under the stress and fatigue of long hours of lambing, standards are allowed to slip. People become insensitive when they are tired and for this reason I believe everything should be done to ensure that lambing is completed as quickly as possible. To have two distinct lambings, one early, one late, which have a tendency to run into each other, is a great mistake. Also, it really does pay to have adequate extra help, and I am much in favour of that being a woman. Her maternal instincts fit her especially well for the work in the lambing pens.

These are general remarks which ought to be almost superfluous.

Of the various diseases which are particularly associated with lambing there is one which is a potential killer on the grand scale. This is the condition sometimes known as 'Watery Mouth' and caused by *E. coli.* The same organism causes considerable losses with young pigs and is definitely a disease associated with intensification. It rarely occurs at the beginning of lambing, but appears as time goes on. Obviously infection builds up in the lambing pens and it is a counsel of perfection that once the disease is seen, those pens should be abandoned. This is easier said than done particularly with inwintered sheep, where the same lambing quarters will be used every year. The symptoms are that the lamb becomes listless, and the mouth will be wet with regurgitated stomach contents, hence the name. Scouring is not necessarily present. Death can occur very rapidly, and with an outbreak there is always the possibility that it may spread very rapidly and get out of control. Cases usually occur in the first two days after birth, but it is by no means uncommon in lambs that have been turned out to grass. It is not easy to understand just why this should happen.

What can be done to prevent *E. coli* getting a grip? One of the problems is that there are a great number of different strains of the organism which make it very difficult indeed to make a vaccine that is any way effective. One company does, in fact, market a combined pasteurella pneumonia/*E. coli* vaccine which we have used for a number of years. Administered four weeks before lambing, it does seem to give some protection, but it is by no means the complete answer. A matter of management which is of the greatest importance is to ensure that the new-born lamb gets its fill of colostrum as soon as possible. This is the best method of all of protecting the lamb, and there is no doubt that of those lambs that do develop the disease, the majority have been short of colostrum.

Treatment of affected cases must be carried out immediately, and is by an oral antibiotic, together with an antibiotic injection. This is effective if done soon enough, but it may well have to be repeated. It is fair to say that we have lived with *E. coli* at Worlaby for a good number of years now. We are obliged to use the same lambing pens year after year and so we are perhaps particularly susceptible. Our routine is that the buildings are scrupulously disinfected during the off-season, and during lambing we do everything possible to keep the pens clean. The ewes are vaccinated, and the shepherds are very alive to the need to keep an eagle eye open for the development of any symptoms. Our losses are very low indeed, and I believe that we can keep it that way.

Of the other diseases associated with lambing, there are mainly two – those where the infection enters via the navel, such as joint-ill and arthritis, and mastitis in the ewe. For the former, an immediate disinfection of the navel cord with an antibiotic spray should be routine. Mastitis we have never found to be serious at lambing time, though others appear to have this problem. When it occurs, then the suspicion must be that hygiene in the lambing pens is not what it should be. Our trouble with mastitis comes much later, generally in May.

(9) *Mastitis*

All lactating animals are subject to the range of udder infections which we call mastitis, as any dairy farmer knows only too well. There is no reason to suppose that the ewe should be immune and nor is she. It is almost always caused by bacterial infection and takes two forms – one streptococcal and the other staphylococcal. Both can cause the loss of either or both quarters, which of course ruins the ewe for future breeding. That caused by staphylococci is much the nastier of the two – often called summer mastitis – and can easily cause the death of the ewe if not treated.

As noted above, streptococcal mastitis can occur very soon after lambing, although in my experience that is relatively rare with ewes that are healthy and free from a previous mastitis infection. If the ewe were to be milked by hand, the first symptom would be the appearance of clots in the milk – just the same as with the dairy cow, and the treatment would be the same, an infusion of the quarter with antibiotics. But of course, the ewe is not milked, she is suckled, and so the first warning we get is a swollen quarter and a hungry lamb. By then, almost for certain, it is too late to save the quarter and treatment can only save the ewe for subsequent culling.

The real problems occur when the ewes and lambs are at grass. Not surprisingly it is those ewes which are the best milkers with heavy udders which are most at risk. Close supervision is necessarily very difficult and almost for certain the first symptoms will go undetected. Cold winds undoubtedly make the heavy milking ewe more susceptible, and I can only assume that chilling of the udder predisposes it to infection, which must be around anyway.

The last danger point comes at weaning, again with those ewes that are the best milkers. Taking the lambs away leads to ewes with stocked up udders, often with some milk leaking from the teat end – providing an ideal highway for the bacteria to enter. And of course, flies can be an added hazard, transferring infection from one ewe to another.

What can we do about it? For mastitis is certainly one of the most important reasons for culling – and as we have seen, often the best and most productive ewes in the flock. Which is quite infuriating! The answer is, I am afraid, not very much. Certainly rigorous culling of dry ewes to prevent infected ewes being around at lambing time. Strict hygiene at lambing time. Keeping the heaviest milking ewes in fields that are reasonably sheltered will certainly help.

Perhaps there is something we can do at weaning. The ewes should be brought indoors, away from flies, and for two or three days put on iron rations of straw and water. The objective being to stop the flow of milk as soon as possible. At the same time each quarter can be infused with antibiotics and the teat sealed – a copy of dairy cow management. It is expensive and very few people think it is worth doing. But with good gimmers costing anything up to £100 and cull ewes worth no more than £30, you cannot afford to have too many culled at an early age.

(10) *Deficiency Diseases*

Shortage of certain minerals can cause disease, notably cobalt (pine), copper (swayback), and calcium/phosphorus (rickets). This is almost always associated with certain specific soil types. Where they occur, the conditions are well known locally, and the remedy is generally easy.

(11) *Pneumonia*

The only thing about which we can be really certain is that there seem to be almost as many types and strains of pneumonia as there are breeds of sheep – and that it is an increasingly common form of loss. We used to associate pneumonia either with housing where the ventilation conditions were bad, or with bad weather conditions where sheep were lying on wet ground for long periods. However, we now seem to be in a very different situation. Both in my own experience, and from talking to many other sheep farmers, it appears that pneumonia in one form or another is becoming increasingly common at all times of the year, and certainly not following any one management or climatic pattern. Losses, often in ones and twos, rather than in dramatic numbers, can build up over the year and add up to a very considerable drain on the flock.

By far and away the most important type of pneumonia is that which we know as '*Pasteurella Pneumonia*'. Here again there are many different strains which makes preventative vaccination a very hit and miss affair. For pasteurella vaccines are available, and no doubt they are effective against particular strains. We are in a similar situation here as we are with *E. coli*. Is it, or is it not, worth vaccinating? It probably is, although the cost can be substantial as the cover is short and must be repeated at regular intervals. But it certainly is not a simple situation which can be controlled with a vaccine, as is the case with Pulpy Kidney for example. It is much more complicated than that. Obviously management plays a part, especially from the point of view of reducing stress to a minimum – and I cannot over-emphasise the importance of that.

What I can say with great conviction is that I would like to see much more research effort put into Pasteurellosis which costs us a great deal more than we suspect, both in losses and in lower individual performance.

Another type of pneumonia, causing considerable loss in some parts of the country, is '*Jaagziekte*'. It is caused by a slow-growing virus and there is no test to determine its presence, nor is there any cure. It would be wrong to suggest that it was of any great importance nationally, but that it causes significant losses cannot be denied either, and it seems to be particularly prevalent in the Borders. It is a fairly certain indication that you have got Jaagziekte if you find a ewe dead and, on lifting her up, head down, a lot of liquid flows out of her nostrils. But you will not know if any other sheep are infected and I hope that I am not being unjust when I say that you will get a pretty blank response from the veterinary profession.

Finally, in a book revised and brought up to date in 1985, it would hardly be possible not to mention '*Maedi-Visna*'. It is certainly very fashionable to be worried about M.V. MAFF run an Accredited flock scheme based upon successive negative blood testing, and Shows and Sales run Accredited and Non-Accredited sections. What is it all about? Let me indulge in some personal opinion and, at the same time, try and throw a little light into the dark corners. Maedi-Visna is a complex of two separate conditions, one of which, Maedi, is a form of very slowly developing pneumonia. Visna can show up as a progressive paralysis of the hindquarters, always in older sheep. It is caused by a slow virus and has been reported from many countries where sheep are numerous. The only time when it seems to have caused serious loss was in Iceland in 1939 after an importation of sheep from Germany. However, it is clear that this was an exceptional set of circumstances, in that the Icelandic sheep were housed for at least eight months of the year in conditions of rather less than perfect ventilation, and that the Icelandic sheep population had been totally isolated beforehand, i.e., it was a comparable situation to that of introducing the common cold to a community of Eskimos who had never encountered that infection before, and therefore had no natural immunity.

The losses in Iceland were admittedly high, and the spectre is therefore raised that similar losses might occur in Britain. Hence all the attention given to M.V. I cannot, however, avoid the conclusion that all this is more than somewhat illogical. M.V. has been known, but certainly not been worried about, in France, where it is known as 'La Brouhite', for a long time. It is not regarded as any particular danger and as so many French sheep are housed under conditions of appalling ventilation, I cannot believe that it would not have flared up if it was going to.

Here in Britain it has only been 'recognised' as being present since the mid 1970s. The recognition is the result of a blood test which shows up the presence of antibodies. It is not therefore a direct test for the disease itself. So what, you may say, if symptoms show up subsequently? But curiously, and fortunately, they do not. Despite a very large number of positive tests having been shown up since 1978, as far as I am aware, there have been virtually no physical symptoms, and most certainly no epidemic.

The response to my reasoning, and experience, is – well, you never know what may happen. Of course there is no answer to that – *except*, we do know that we lose very large numbers of sheep annually from Pasteurella pneumonia. I only wish that all the veterinary effort which has been put into chasing Maedi-Visna could be directed with similar enthusiasm against Pasteurella. My humble view is that M.V. is a phantom which, like all ghosts, is scary, whilst Pasteurella is a proven killer.

CONCLUSION

I finish this chapter where I began. It is not a veterinary dictionary. It is a commentary from a sheep producer on some of the main disease problems which he has faced in the course of developing more intensive management. You must read it in that spirit and be sure that you do not believe that it is a complete guide. Consult your local vet; through him make use of your Veterinary Investigation Centre. There is plenty of help available. But one word of caution. Don't let anyone try to persuade you that the only way to have healthy sheep is to have fewer sheep.

Chapter 15

MARKETING THE FINISHED PRODUCT

A. The Producer

EACH PRODUCER must make his own decision as to what form of production is best suited to his farm, with its own particular conditions. Having made his choice, be it for early lamb, late lamb, store lambs or breeding ewe lambs, and then carried out the production of his chosen line to the best of his ability – having done all that, then much can be gained or lost by the way in which he finally puts the finished article on the market. In this chapter, I propose to deal with the final product, the slaughter lamb, from the point of view of the producer and then, further down the chain, at the abattoir and in the world markets.

The British producer has had to live with the realities of the world market, and in particular the influence of cheaply produced New Zealand lamb. The harshness of this reality was greatly softened by the institution in the Agriculture Act of 1947, of the annual determination of guaranteed prices at the February Price Review. Of course, the reality remained in that the level at which prices were set was greatly influenced by the New Zealand trade as well as by the other main objective of successive Governments, that of maintaining just a certain level of sheep production in Britain. The profitability of the sheep industry as a whole was largely determined by these price-fixing decisions. Broadly speaking, the only way in which an individual producer could improve his profitability was by increasing his production efficiency. Marketing skills were redundant. Each individual would get the guaranteed price, either from the market or from a market price made up by the deficiency payment.

This is not the place to describe in detail the operation of the

Deficiency Payment system, which is now past history. I will just say that it had a deadening effect on marketing enterprise. It cut the lines of communication between consumer and producer. Up to certain maximum weight limits, and subject only to the application of fairly loose grading standards, a producer would get paid the same price irrespective of qualities. Indeed, it was completely true that profitability was directly related to selling at high weights. The trade could talk endlessly about the desirability of producing lightweight lambs, and in particular, carcases with little fat cover, but all to no avail. Few producers were silly enough to follow the exhortations of the trade against their own financial interests. This then was the background against which lambs were sold from 1947 right through to the mid 1970s.

Britain's entry into the Common Market should have changed all that. Market demand, and in particular the Continental market, should have been the determinant of reward to the producer. How sad it is that, in 1985, I should be writing 'should have'. The hope and the optimism that, at long last, real quality would be given its just reward, have faded away. The reason is quite clear – the Common Market has never been allowed to operate for sheep meat. Instead we have national policies – or to be more precise, we have a British national policy and a French national policy. It is easy to understand why, even if the underlying causes are not exactly based on logic. The presence of New Zealand, and our attachment to her, still hangs over the market. The French see no good reason why their sheep industry – equally vital to their poorer areas, as ours is – should be sacrificed just to allow British lamb to come in whilst the British eat New Zealand lamb. I can understand and sympathise with that point of view.

The end result is that the French market is protected against imports from Britain by what has come to be known as the 'claw back'. That is to say, any subsidy which is paid in Britain on lamb is added back to the export price. In consequence, for much of the year, lamb cannot be exported to the lucrative French market at a profit.

Here in Britain, we have the Variable Premium scheme which we might just as well have called the Deficiency Payment scheme for that is precisely what it is. It has the same dead hand resting on the market obscuring the quality cash message coming back from the consumer.

QUALITY LAMB

It is one thing to say that the trade wants, and is prepared to pay for quality; it is quite another to know exactly what to produce. What weight of carcase should I aim for? What level of 'finish' should I put on my lambs? What breed of ram should I use to produce these slaughter lambs? The answers to this type of question are very clearly defined in the pig industry, but we sheep producers are a long way away from such a goal. It is fair to say that it is impossible to make any sensible generalisations which apply to the whole country. Notwithstanding the influence of the export trade, which is, despite everything, strong in some areas where certain abattoirs have made it their speciality, there are still important regional differences in Britain. And any producer would be stupid to ignore these.

This is all very unsatisfactory, and perhaps the gist of this whole chapter will be that we really must introduce some order into what is still a chaotic situation. What one can say for sure is that the consumer of lamb, of whatever nationality, does not like fat and is not prepared to pay for it. Even more, that average consumer is on the way to being convinced that animal fat is a health danger and is not prepared to take that risk. Particularly when turkey joints, even if somewhat tasteless, are fat free to say nothing about being cheaper and enormously more convenient both to cook and to carve. There is scope for discussion about the correct weight and most suitable conformation, but if we wish the housewife to buy our lamb, there must be no argument about fat. Marketing quality lamb ought to be all about marketing lean lamb.

SELLING: AUCTION OR DEADWEIGHT?

We are selling meat, in this case lamb meat. That being so, then there can be no denying the logic that the only way to judge its quality is when it is in the form of meat, that is, hanging up on the hook in the slaughterhouse. At that point, there can be no argument about weight, or fatness, or conformation. It is there to be seen and measured and paid for accordingly.

Why, then, in the face of this logic, is it that the majority of producers continue to sell their lambs in the fatstock auctions; and the bulk of procurement for the abattoirs is done by buyers operating in these auctions? And why is it that we in Britain continue to do it this way when such widely differing countries as France and New Zealand are entirely committed to deadweight marketing?

It seems to me that this is a question of fundamental importance. I cannot see how we can bring about real progress in improving the quality of British lamb whilst the majority are sold live in markets. It is just not possible, however skilled the buyer may be, to make sensible judgements of carcase quality by visual assessment in the conditions of a livestock auction. Without that judgement being passed back to the producer in price differentials, no progress is possible.

In trying to answer the question, one could be completely cynical and say that the average fatstock producer enjoyed his day out at market. The alternative of staying at home doing some work looks pretty unattractive! Studying market trends in the bar of the Red Lion is a form of research that we can all enjoy. However, it is not the purpose of this book to be either cynical or amusing at anyone's expense, but to be constructive, so we must look further afield

There is one very good reason, and as always we return to money. Farmers react to financial incentive as quickly as any class of businessman I have ever come across. The fact is that the auction market has consistently given the producer a higher return than marketing deadweight. It renders no service to the industry at all to shut one's eyes to this fact, as some try to do. The fact that this is true does not alter the fact that it is crazy.

The next buyer in the chain is the abattoir-cum-wholesaler. The lamb will finish there in any case, whether it goes there direct or via a market. Putting an extra link in the chain ought to increase cost and therefore reduce return, instead of the reverse; quite apart from the damage which is done to the lamb by subjecting it to stress of the market and that of an extra journey.

The advantage of the market is that it is an assembly point where relatively large numbers of lambs are gathered together at the producer's expense. A buyer can make his selection from this assembled collection, and do it all in the course of a few hours. By contrast, he would have to drive many miles and see many small lots of sheep on different farms in order to be able to buy the same numbers. For a buyer who may well have to supply at least 3,000 lambs a week to meet his abattoir's requirements, it is physically the only way he can do the job. In consequence, pushed along by market competition from his fellow buyers, he will pay something for the convenience.

So that is the advantage from the buyer's point of view. So far as the producer is concerned, he has the feeling that he can always take his stock back home if he does not like the price; rare though that may be. Of greater significance, there is the fact that the market is undoubtedly the best way of disposing of his poorer quality stock. To a greater or lesser extent, all of us produce sheep of which we cannot be very proud. There will always be tail-end lambs and cull ewes to dispose of. Very often the top-class abattoirs, aiming at the supermarkets and the export trade, are not very keen to take this class of animal. But they are there to sell and there are those in the trade who wish to buy them. The auction market provides the means of bringing the two together.

But is it the only practical way to meet the abattoir's procurement needs and at the same time to give the producer a better return? I believe that it would be both pessimistic and unrealistic to accept that this is so. Both producer and purchaser have a part to play in finding something better.

QUALITY LAMB GROUPS

These groups have been with us for some long time now, all started by enthusiasts who have been inspired by the need to market better. All have followed, more or less, the same pattern. They have been made up of regional groupings of producers, generally under a cooperative umbrella, who have agreed to contract all or part of their lamb production to the group. The group in turn has offered a contract to supply an abattoir with an agreed number of lambs over a defined period of time. In other words, it has been a planned procurement operation, and in return for this service, the group has sought to obtain a premium from the purchaser. In addition, most groups have sought to impose some sort of quality control, though in most cases this has not amounted to much more than a control of liveweight.

It is fair to say that, despite the enthusiasm of their promoters, and significant grant-aid from the Central Council for Agricultural Cooperation, group marketing of lambs has not really caught on. Again, we come back to money. The premiums received have not been sufficient to do much more than cover the overheads of running the group. And it is very easy indeed for the overheads of such an operation to run away and get out of control. What one can say is that they have been an honest attempt to provide procurement for deadweight marketing in order to counter the assembling advantage of the fatstock auction. They have not yet, by the mid 1980s, been able to go far enough down the road of providing the consistently high quality which would enable the abattoirs to pay them a significantly high premium. That is not entirely, or even mainly, the fault of the group. Much more of the blame, as always, must rest on the deadening effect of the Variable Premium Scheme.

B. The Trade

Let us consider for a moment the state of the slaughtering section of the industry as it faces the last decade and a half of the century. That it has gone through a painfull revolution is beyond doubt. The number of slaughterhouses in Great Britain has fallen from 1,890 in 1971-2 to 1,062 in 1981-2. These figures seem to suggest a sensible rationalisation, eliminating the outdated and concentrating capacity in modern, labour-saving plants. To an extent, this is true, but it has certainly not given us a lean, fit and prosperous slaughtering industry. Numbers of abattoirs may have fallen, but capacity has actually increased to the extent that over-capacity is a serious economic problem. Add to that the problems of the economic recession of the early 1980s, and the disappointingly poor export trading, and you have a pretty bleak picture. If anything, the situation has got worse since March 1982 with a number of the most modern plants, that have led the way with new techniques and innovations, reported as being in serious financial difficulties. The paradox is that it is often those very abattoirs which technically are in the very forefront of development that are in most danger, quite simply due to the huge cost of modernisation. The little local abattoir with poor facilities and only just conforming to the hygiene regulations, but no load of expensive borrowing to service, is often reasonably profitable.

So there is a desperate problem of over-capacity. Surely though that problem will sort itself out even if some companies do not survive. The modern abattoirs will remain and will no doubt be bought at a price that allows for profitable operation. But the problem is more complex than that as there is actually a shortage of the right type of abattoir. In March 1982, only 71 were listed as EEC Approved Abattoirs. That is to say that under present EEC regulations, the rest are not permitted to slaughter for the export trade, only for the home market. What an admission to have to make that our slaughterhouses are only fit for ourselves, not for the foreigners! But that is the truth of the matter.

So the industry has a fundamental structural problem. Far too many abattoirs – most people would agree that 400, strategically situated both in relation to production and to the markets, would be sufficient. And a crying need for further investment in modern technology which is extremely expensive. All this against the background of a poor profit record. To say that the problems are considerable is something of an understatement.

As if all this were not enough, there is the problem of the external influences of the export/import trade. I have mentioned already the harmful effect of the so-called 'claw back' on lamb exports. Then there is the temptation for Governments to play the Common Market and in particular the 'green currencies' which artificially distort trading patterns. So we get commercial advantage given to imports in carcase form from Ireland, Holland and Denmark and the frustrating thing is that it has little or nothing to do with normal trade patterns or the efficiency of production. Even an efficient industry, fully equipped and in top health, would have a job to cope with that.

So as we producers look beyond our farm gate, we see our primary purchasers in a state of disarray: struggling to pay today's bills and with little energy left for constructive thought for the morrow.

However, it would be going far too far to say that these problems are insoluble. Given the political will, allied to some vision in Whitehall that looks just a little further than next week's shop prices, the economic solution will be found. Indeed, it is being found by those few individuals who have both courage and vision. Their task would be so much easier if they were not so badly handicapped by the unfair competition from subsidised imports. We must hope that this will change. Then, as these abattoirs have an increasing influence on the marketing of our lamb, we as producers must concentrate more on meeting their requirements.

THE NEEDS OF A MODERN ABATTOIR

All abattoirs will be dealing with more than one species, but I shall stick to lamb in this consideration of their management. In doing so, I am indebted to my friends in Canvin International of Cardington, Bedford, with whom we work, and who are one of the leaders in the slaughtering and meat wholesaling world.

What do they need from us?

Procurement. First of all, sheer numbers – of the order of 5,000 lambs a week will be needed to keep their sheep line going at full capacity. Quite a daunting prospect, to keep that number flowing in regularly. Not only is the problem different in terms of sheer quantity compared with a small low capacity plant wanting say one-tenth of that number, but it also greatly increases the geographical spread of the procurement operation. But procurement is about more than sheer quantity – it is also all about quality. For the lamb is only bought in order to be sold again, and that sale will be made to a tight quality specification. So the buyer will start his week in the knowledge that he must buy a large number of lambs, and that each lot must conform as closely as possible to the requirements of different orders coming in from the various trade outlets. As I have pointed out earlier, the bulk of this procurement is still done in the fatstock auctions. Not only does this lead to continuing uncertainty about the supply, but meeting the quality specifications is extremely difficult.

The quality market at home and abroad. Any big abattoir must be seeking to supply big outlets. Dribs and drabs of sales to small customers cannot be part of the business. As the slaughtering industry moves to a smaller number of bigger units, so will the move away from the small butcher to the supermarket and multiple chains accelerate. The change matched exactly by the evolution in High Street selling in favour of centralised cutting, so that the butcher's shop becomes merely a sales counter.

The big outlets demand and expect to get a specified product. The specification may well vary from one outlet to another, but for any one, it will be precise. If this is true for the home market, then it is even more so for the export trade. Selling across the Channel to the high priced markets in France and elsewhere, may well be very lucrative, but it is also very demanding. It is quite plainly commercial suicide to send lamb carcases to a Continental buyer that do not meet his requirements. He will just not be interested at any price, and no further business will be coming your way.

So imagine, if you can, the stream of live lambs coming in to the plant, perhaps 1,000 a day from different markets, different farms, different areas and, worst of all, from goodness knows how many different breeds and crosses. And picture, going out of the same plant, lorry loads of carcases, each lot like peas in a pod and conforming precisely to the customer's specification. Some miracle has to have happened in between! What has happened, of course, is that there is a considerable percentage of discards, carcases which are too big, too fat, too lean—which fit no specification.

Problem of Unwanted Carcases

These discards do not necessarily have to be bad lambs. They may be, but they can be discards simply because the abattoir management has not got a trade for them. If they had *known* that they were coming, and if those in each catergory were in sufficiently large numbers, then they could have, and would have been pleased to find, the right outlet for them. As it is, they are something to get rid of, something they have no planned sale for. And what generally happens is that they are put on Smithfield Market and sold for what they will fetch. Have you ever wondered why, if you listen to the early morning market reports on the BBC, the trade reported from Smithfield is usually way beneath what you are asking ex-farm? That is the reason why. Markets like Smithfield have become the 'dustbin' into which these unwanted carcases are put for disposal sale.

The main point to appreciate is that these unwanted carcases can be, and frequently are, an alarming high proportion of the lambs coming into the abattoir. At some times of the year it can be as high as 30 per cent, although a really skilled buyer may succeed in keeping it nearer 10 per cent. All are bought at the same price, but this proportion has to be disposed of on a low price market. The manager of an efficient modern abattoir has almost certainly succeeded in getting a significant premium over the market for his quality sales to the big outlets, be they at home or abroad. But his disposal of the unwanted lamb represents an overhead cost which often cancels out the premium obtained. Make no mistake about it, this is a serious problem. If the solution can be found, then not only will the abattoir be more profitable, but money will be there to pay quality premiums back to the producer.

Is this all the fault of the buyer who is unable to come within acceptable limits of meeting the specification given to him each week? Given that he has to buy in the auction markets, then it will be a very skilled buyer indeed who will get 90 per cent of his purchases falling into the right bracket. An almost impossible job, faced with lines of pens of sheep, none of them graded in any sense of the word. He cannot weigh lambs individually, a quick feel of a few backs, a hurried bid, and on to the next pen. He relies on his skill and experience of visual judgement. The miracle is that he does as well as, in fact, he does.

But it is not a very logical way of providing for the nearly insatiable hunger of his abattoir boss, who must keep up to capacity as well as keeping his customers happy.

Why, then, not use the Quality Lamb Groups? Do the procurement direct from the farm via a group contact. The answer is that an increasing number are doing just that. What we need to debate is how this move can be accelerated and, more important, how it can be improved.

Think for a moment about the pig industry. Thirty years ago there were similar problems. A multiplicity of breeds and crosses, fattened under all sorts of different systems, and most of the pigs marketed in the auctions. Hardly any need to underline the tremendous change that has come about. The British bacon pig is now of uniform and very high quality. The fact that the bacon and ham business is in such a parlous economic plight is certainly not the fault of either production or procurement. What has brought this change about, and has it any relevance to the present sheep muddle? The answer is simple. A precise spelling out to the producer of exactly what the bacon factory requires, allied to the scale of payment which rewards the good and penalises the bad.

It will immediately cross your mind that pigs and sheep are so dissimilar that any comparisons between them are not very helpful. Pigs after all are kept in houses where the methods of production can be standardised, be the farm in North Eastern Scotland or in Cornwall. That can never happen with sheep, which must remain essentially exploiters of forage. A certain multiplicity of breed must therefore remain in order to cope with differing climatic and soil conditions.

Of course, I accept that. What I do not accept is that this should be used as an excuse for doing nothing.

Whose Responsibility?

In a way, this is a chicken and egg situation. Let me repeat the essentials. A precise spelling out of what the market requires plus payment based on premium and penalty.

It would seem from that to be quite clear that the initiative must come from the purchasing abattoir. The problem is that the average lot of lambs from the average producer is almost certain to attract as much penalty as premium. A premium scheme sounds fine in theory because every optimistic individual, which accounts for us all, will assume that his lambs will therefore in total fetch more money. The reality is often rather different. Faced with reality, the producer sees no incentive to continue and returns to the auction, where his poorer quality lambs get swallowed up in the whole. This is particularly true in times when prices are firm and the market is under-supplied with lambs. We have had this situation right through the latter half of the 1970s under the continued influence of inflation and an increasing export market. Some of the abattoirs operating ex-farm procurement, on a graded price scale, have been forced on occasions to revert to market buying simply because their supplies were being diverted to the markets.

Nevertheless, taking all these difficulties into account, the first initiative must come from the abattoir. Farmers respond, as I have said on countless occasions, to economic incentive. Preaching is a diversion best kept for Sunday mornings.

What Quality?

The starting point has to be some much clearer definition of what the market wants. At the present, the only message coming through with any clarity to the producer is that carcases must be lean. And even the message, about which there should be no doubt whatsoever, is blurred more than somewhat by the grading standards applied by Ministry graders at the markets. These favour the 'well-finished' lamb which almost invariably means a lamb carrying too much fat.

What weight of lamb carcase does this trade want? Does it lie in the 16–20 kg deadweight range, smaller in the summer, heavier in the winter? I, as a producer of carcase lamb and as a breeder of my Meatlinc sheep whose sole purpose it is to sire carcase lambs, have to say that in 1985 nobody is prepared to tell me. Not for this year, never mind what I should be producing in five to ten years' time. And bearing in mind what an excruciatingly slow business sheep breeding and improvement is, I need to be looking that far ahead today in my selection.

I am bound to say that the trade is in no position to criticise the sheep breeder until such time as a clear statement is forthcoming about what the trade really wants.

CARCASE CLASSIFICATION

I am talking about quality. I have already posed the question – what quality? And said that there is need for a clear statement. But in what language?

Weight, yes this is easy. But how lean is lean, and what is too fat? The pig breeders have got that with backfat measurement. And conformation, what do you mean by that? Do you mean heavy fleshing in the hind leg? Or do you mean more than that, to include a well-developed eye muscle in the chop? To tell me that the live lamb should 'handle well, be firm across the loin, with a well-finished dock', is all very artistic but it does not get me very far.

This is what carcase classification is all about. A description in a common language which all understand, and which means the same thing all over Britain, and eventually, hopefully, all over Europe. The MLC Sheep Carcase Classification scheme uses a descriptive grid so that a trained classifier can place an individual carcase in its proper category.

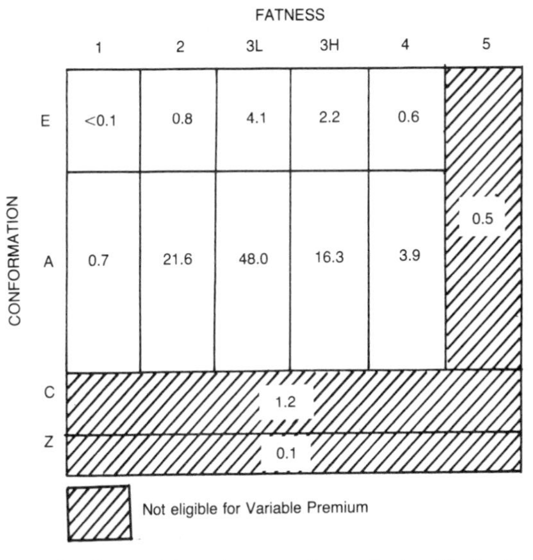

Fig 6. Classification grid and proportion of classified classes in each category, 1983

The placing of a carcase in a certain category, say for example at 2E, at a deadweight of 18.5 kg, tells me something about that carcase. And it tells me that irrespective of breed or time of year or state of the market. It does not place any value on the carcase, it merely describes it.

Classification is then a language, a common tongue. The difficulty lies in obtaining consistency in annunciations of that language over the country as a whole, applied by different classifiers who are, after all, individuals with different judgements. However the admission that there can be some variations despite uniformity of training and constant checking, should not be allowed to invalidate the process. For a

30. Typical auction sale of Halfbred ewe lambs.

31. Five lamb carcases illustrating each of the five fat grades in the MLC carcase classification scheme.

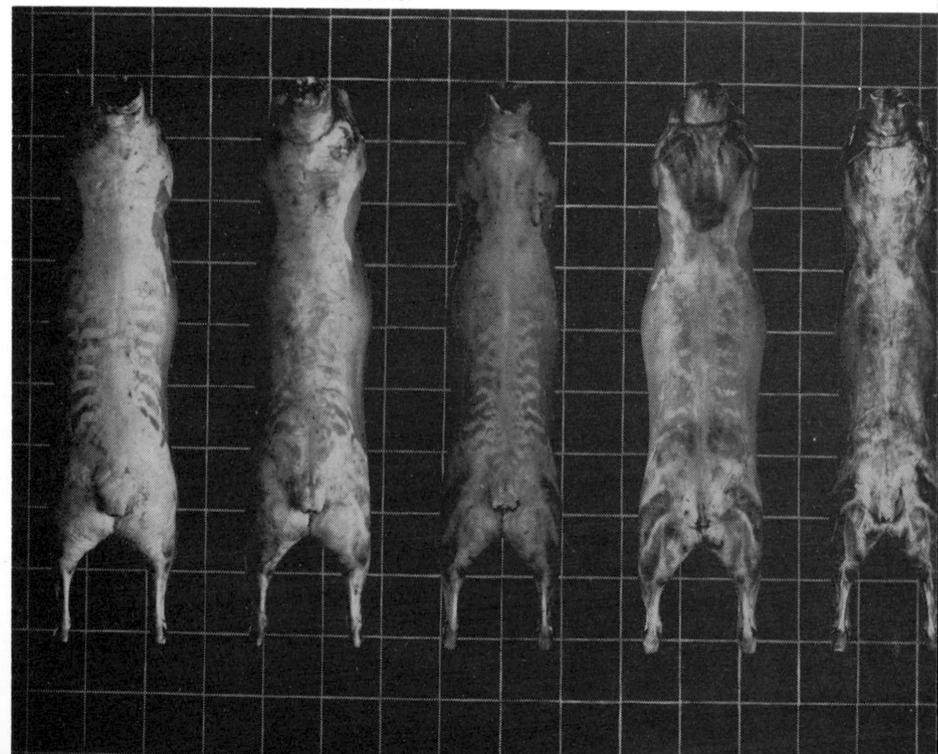

meaningful description we must have if we are to make progress either in breed improvement or in the day-to-day selection of lamb for slaughter.

It was for us at Worlaby, both for me and for my shepherds, an enlightening experience to move from entirely market selling to direct selling to an abattoir, using carcase classification. What we had previously thought of as reasonably well finished were found to be over fat, classifying at 4 rather than 3. We were keeping our lambs ten days too long, spending money and time putting on fat which the butcher subsequently had to trim off before he could sell the joint to the housewife. We were all losing from start to finish

Classification, then, is a valuable means of education in the selection of lambs for market. It should be used much more as a tool in livestock improvement schemes. With some refinements, it should be the basis for description on which premium and penalty pricing schemes are based.

A CONSTRUCTIVE PROPOSITION

Whenever you look in agriculture, at those examples where high and consistent quality is achieved, and marketing is successful – bacon, sugar-beet, frozen vegetables, particularly the pea – one thing stands out as a common factor. The primary purchaser dominates production. And I do mean dominate.

Let me take the frozen pea as an example. A given factory will cover a certain area of production. The major freezing companies will have factories situated in the main production regions of the country. Each will process a range of different vegetables and probably fish as well in order to have capacity production throughout the year. But for its peas, each factory will have a membership of growers on long-term contracts who do exactly as they are told. They grow the varieties provided by the factory, and they drill on a specified date and harvest, again, under instruction. Of course, it goes without saying that problems arise, from vagaries in the weather, if nothing else. But the end result is efficient marketing of a uniformly high quality product, with very little indeed going into the low grade markets. It is worth noting too that the major companies are not only involved in production, processing and marketing, but also in plant breeding and plant health.

Now pea growers, as a bunch of farmers, are every bit as independent minded as sheep men. The point is that they and the processors find that it *pays* to do it that way, together. Nor indeed do the processors have to be, by definition, big industrial companies such as Unilever. One of the major influences in the vegetable processing scene is a farmers cooperative company.

I have gone on at length about peas because I believe that there is a pattern, an example, here that the big abattoirs ought to follow.

The essentials are:

(a) They have to get involved in production, both by contract and by putting money into it.
(b) Each will have to do this in different parts of the country so as to get as near to year-round production as possible.
(c) It must pay the good producer handsomely. This means disciplining out of the indifferent and generously rewarding the good.
(d) It means a deep commitment to animal breeding, although this probably need go no further than the final sire line. Whether one single breed of meat ram can do this country wide is open to question. Certainly no more than two, one early and one late maturing. But each should be selected centrally so that there is a real genetic uniformity in the breed which will provide 50 per cent of the make up of the final lamb product.
(e) It means a field service by men who are both husbandry advisers and procurement officers.

Is this an impossible pipe-dream for an abattoir such as I have been discussing – 5,000 lambs a week, say 200,000 lambs a year? At a delivered abattoir value of, say, £40 a head, that is an annual value of £8 million. I would have thought that it was worth a lot of effort to get that right! Improvement in the sheep world has been desperately slow in coming, and nowhere more so than with carcase quality. All this despite the tremendous effort put into shows and showing, into the winning of breed championships and breeding from such champions. We have to break away from all that, and for no one are the rewards greater than for the big abattoir/wholesalers whose future livelihood will depend on this quality. Surely they dare not leave it all to chance?

THE REALITY OF THE MARKET

I wrote the preceding section in 1978 and, apart from updating the financial values to 1987 levels, I see no reason to change a word of it. It still seems to me to be the only way by which we shall bring about dynamic progress. It has not happened and there is no sign that any such changes are likely.

Let us pause and consider what are the realities in the only market that matters – that is to say the butcher's shop, be it supermarket or individual.

Consumption of lamb in the UK has fallen dramatically and is still falling: from 423,000 tonnes in 1982, to 407,000 tonnes in 1984, to 378,000 tonnes in 1987. This is happening despite the fact that lamb is effectively a subsidised meat to the housewife. If the Variable Premium Scheme has any value at all, it surely ought to be that it is *selling* the meat via a producer/consumer subsidy rather than putting it into intervention. With a large proportion of the producer's return coming from Brussels (on occasions up to half), it would be reasonable to expect to see lamb dominating the market. Instead, consumer research tells us that the housewife, particularly if she is under 35, has reservations about lamb. She considers it to be fatty, wasteful, lacking in versatility, difficult to carve and poor value for money. Not really the best quotations to put into a lamb promotion campaign!

Against such a market background, one might expect to see signs of a declining production. Far from it. In fact, sheep numbers have been increasing steadily. Ewe numbers have gone up by 7.8 per cent from December 1980 to December 1983 – a reflection of the increased returns of producers coming from the British version of the CAP for sheep meat.

I am bound to ask – what sort of reality is this?

Something has to be wrong somewhere – and it is no good blaming the Common Market (the standard excuse of the British for any agricultural problem!) for the Sheep Meat Regime is our invention. Nor is it any good saying that all that is necessary is some expensive TV promotion. All the promotion in the world will fail if the customer does not like the product.

So what can I add to my 'Constructive Proposition'? The message coming from the market is quite clear – lamb *must* be lean, fat is out. Yet the cash message coming to the producer, i.e. the combination of price plus variable premium subsidy, tells him that the most profitable lamb to produce is at 23 kg deadweight and Fat Class 4 is fine but be careful that you do not go into Fat Class 5 because then you will lose the subsidy. But Fat Class 4 is far too fat for the housewife – ask any wholesaler selling to Sainsbury, Marks & Spencer or whoever, and they will tell you that they want, and *cannot get enough of,* Fat Class 2 and 3L.

It should be quite clear that the devil of the piece is the way we have our variation of the Sheep Meat Regime structured. It is laudable that it markets meat at a lower price – but it is utterly to be condemned that it positively encourages lambs which are too heavy and too fat. There is nothing wrong with the scheme in principle; it is only the emphasis which is wrong, so for goodness sake let us change it.

So to be specific – let us cut out Fat Class 4 from the subsidy; and the price penalty will then be such that producers will stop producing them. But we have to go a bit further than that. Where lambs are sold in a fatstock auction, they are graded for subsidy by MAFF graders. Whilst it is practical to expect these men to distinguish, under all the pressures of a market, the Fat Class 5 lambs, there is no way that it is reasonable to expect them to distinguish between Fat Class 3 – into subsidy – and Fat Class 4 – out of subsidy. Only carcase grading can do that via the already established carcase classification system – why on earth not use it and pay subsidy at the abattoir point irrespective of whether the lambs are sold in the market or direct on a deadweight basis. There would be no difficulty in arranging this either:

- By obliging the purchaser at the fatstock auction to bid the whole price, i.e. the purchaser taking the subsidy at the abattoir.

or

- By retaining identification of the lambs through to slaughter so that the producer received the subsidy.

The first of these two alternatives would be by far the simplest. Both would have the great advantage of doing away with liveweight grading – which is a near impossible job anyway – whilst retaining the option of selling at live auction which many producers prefer.

All this seems so logical to me that I wonder why we do not do it. Why does the NFU oppose such a change? I only wish I knew. What I do know is that by burying our heads in the sand, we are losing our market – and in the real world, that can only have one eventual result.

THE WORLD MARKET

I started off this chapter by saying that British producers had become accustomed to living with the realities of the world market. By this I was referring, of course, to the dominance of New Zealand. Although this was complete, it did have the virtue of being simple. We produced some 55 per cent of our total consumption of lamb and mutton, and New Zealand supplied the rest. Furthermore, the imported frozen lamb fitted in well with the seasonality of British production filling in the gap during winter and early spring. It goes without saying that the quality of these imports was very good, being selected to a high standard by the New Zealand Meat Board. So much so that the description 'Canterbury Lamb' passed into the English language as an accepted description of top quality succulent lamb. A tribute indeed to the promotion abilities of the New Zealanders, who could so succeed with a product which, of course, had to be frozen to withstand the journey. Be this as it may, marketing of British lamb consisted of selling it to the British public at a price which always had to have a relationship to that of the imports. There was certainly no aggression evident in our marketing, exports were non-existent and we were content to sit back and accept that our national flock was at a certain level which suited the country and was unlikely to be changed.

This whole approach has completely altered since our entry into the EEC. New Zealand still exists of course, and whilst our imports have fallen from 255,000 tonnes in 1970 to 184,500 tonnes in 1983, that is still a considerable figure. And indeed, total sheep and sheep meat imports into the EEC in 1983 amounted to 230,449 tonnes. Exports to non EEC countries are relatively small and amounted to only 4,241 tonnes in 1983. So the community of 12 is still a major importer of sheep meat and is only 74 per cent self sufficient. Imports are therefore certain to remain a feature of the EEC lamb market for the foreseeable future, even though coming in with the considerable penalty of a 20 per cent ad valorem import duty.

However, the voice of the consumer in Europe gets more powerful by the day as the proportion occupied by the agricultural vote in each of the Community countries declines. Consumer taste is changing at the same time. Whereas the French housewife has had the reputation of being a daily rather than a weekly shopper, interested only in purchasing fresh what she needs immediately, and hence buying from the corner shop, she is now adopting the habits of her American and British counterparts. A weekly visit to the hypermarket. The purchase of a deep freeze. This is becoming the pattern. Add to this inflation plus the desire to spend money on leisure rather than food, and we are but a short step from the time when the Continental housewife will demand that she has free access to the cheaper lamb which can come from overseas.

It is my view, therefore, that, in the long term at any rate, the EEC lamb market is more likely to settle down along the lines of the British model rather than that of a high priced, luxury meat, the production of which is primarily the provision of community farmers.

That however is for the future. The potential, if only we have the courage to grasp it, is exciting. Lamb is one of the very few products where the community is in deficit production. By contrast with milk and cereals where the future seems to hold no hope other than restriction, lamb ought to be looking hard for growth. Growth not only to fill Community needs and cut out imports but growth into the export market as well. And the British sheep industry ought to be in the forefront. 'Ought' I say, but will it? Only if we grasp the nettle of disciplining ourselves to produce quality.

Our fellow member countries of the EEC remain our most attractive market. This, despite the fact of French barriers to free trade within the Community which must surely disappear in the near future. British lamb has established a significant foothold in France and also in Germany, Holland and Belgium. The prospect for expanding this market, by way of promotion to change people's eating habits, plus aggressive marketing, must be great.

The market place is, however, wider than Europe. British lamb is now sold by the 'plane load in the Arab countries of the Middle East. The quantities are not yet great, but the potential, albeit in competition with the New Zealanders and the Australians, is considerable

So the change is profound. From accepting the realities of living with cheap New Zealand imports to going out into the world markets to capture trade as an exporter. From passive acceptance of a static national flock to the real hope that a significant expansion is a real possibility.

Are we well equipped to take advantage of the potential? To counter the optimism and the enthusiasm, the answer must be 'no'. So few of our slaughterhouses are up to standard. So many are out of date and near bankrupt. A few, but far too few, of our wholesalers are mentally attuned to the demands and the toughness that are part and parcel of developing new markets. Too many prefer to sit on their backsides bemoaning the fact that life is hard and meat is too dear. Too little influence is coming back down the pipeline to the producer to cajole and encourage him to produce what the market wants.

But this is the dark side. The bright side is that there is a handful of wholesalers whose spirit of enterprise is typical of the British entrepreneurism at its best. There is a wonderful base in the British flock on which we can build increasing production of the best quality lamb at competitive production costs. What we do need is a far greater awareness from Government that British lamb is an export asset of considerable value.

Chapter 16

ECONOMICS

I AM not an economist, far from it. I do not even particularly enjoy sitting at my desk juggling with figures; unless, that is, they happen to be particularly good ones. So let me start by saying that this chapter is very definitely not going to be a manual of economic standards, targets and performances. Apart from the fact that I should do it badly, I would be setting out to provide information which is very comprehensively covered elsewhere. I have great admiration for the work of the Sheep Improvement Department of the MLC, and in this context, the economic information which is published and is based upon the commercial recording schemes of the MLC. There is a wealth of information here, clearly analysed according to different systems and which is available to anyone who cares to use it. In particular I would recommend my readers to study *The Sheep Yearbook* published annually by the Economics, Livestock and Marketing Services of the MLC available from Queensway House, Bletchley, Milton Keynes, MK2 2EF.

When I think back to my early days at Worlaby, a time when sheep breeding and management was in a state of total depression, the thing that strikes me most was the almost complete lack of any constructive thinking and, worst of all, of real information, particularly economic information. Apart from the general statement that a higher lambing percentage was the way to higher profit, there was nothing. A few costings from University Economics Departments, but they were sketchy indeed. Nothing that could be said to be an analysis of what were the economic facts of the industry country wide. Certainly not any analysis of what could be done to improve profits. The fact that we have much of this information today is due to MLC's work and we have cause to be grateful for it.

So that is what I am not going to do. What I do propose to do is to discuss the various factors of cost and return which affect profit. Perhaps more particularly those to which we need to pay attention if we are going to pull sheep production up out of the mundane into the higher reaches of real profitability. Whereas during the years since the war and right up until the mid-70s, sheep were very definitely in the poor relation category, the transformation which has come about following Britain's entry into the EEC has put them into a different league altogether. Furthermore, sheep meat being one of the very few products which are in under-supply in the Community as a whole, the future prospect both for import substitution and for increasing our own share of the market are bright.

Thus I start this chapter on the economics of sheep, be they on the hill or in the lowlands, with the conviction that we are talking of an enterprise with good profit potential. And one which merits serious study and attention to detail, such as you would expect to give to a herd of dairy cows or the potato crop, and not only merits such study but demands it if success is to be achieved.

THE STARTING POINT

As I have said, when talking about the development of the flock at Worlaby, there is simply no way that you can transform the ewe into an intensive production unit in her own right. Nature has equipped her with two teats and a five-month gestation period, and it is my view that it is far better to accept these limitations and work within them. Yet, curiously, and I suppose it is a reflection of the stubborn tenacity of such men, the researchers have concentrated much of their effort into trying to overcome these limitations (a) by breeding so as to improve the capacity to produce litters of three or four lambs, and (b) by hormonal treatment to induce out-of-season conception so as to give perhaps three lambings in two years. That these things are technically possible is not in doubt. No one has so far shown that they are economically worthwhile.

Yet if we are not prepared to accept the argument that the greatest overhead cost in a sheep enterprise is that of keeping the ewe for the whole year, and that the only way to reduce the proportion which this overhead cost bears on return is by maximising those returns per individual ewe by producing more lambs, then we have to find other ways of increasing output. For on high cost land, and at today's values virtually all land except the barest mountain is relatively very high cost land, high output is essential for profit. That remains true whether we are talking about a hill farm with a stocking rate of 1 ewe per hectare and a lambing percentage of 90; or a lowland farm with a potential of 15 ewes per hectare at 175 per cent.

OUTPUT

In a lowland situation, and taking 1987 figures, the sheep flock will be in competiton with such crops as wheat and barley; or more probably, because we are talking about alternative crops, with oilseed rape and sugar beet.

Forgetting about costs for the moment, let us look at a reasonable level of gross sales that can be expected from average yields of these crops.

Wheat	7 tonnes/ha	@	£110	=	£770/hectare
Barley	6·5 tonnes/ha	@	£100	=	£650/hectare
Sugar Beet	38 tonnes/ha	@	£26	=	£988/hectare
Oilseed Rape	3·4 tonnes/ha	@	£220	=	£748/hectare

Perhaps it is a reflection of my own arable/sheep background that I should take these crops as my basis for establishing what the sheep flock has to achieve. It is the environment in which I live. But just as important is the fact that if we are to see any significant expansion in the national flock, a good deal of it could come in the lowlands. And if it did, then it would have to come against the background of such figures.

Another general point that I should make about these figures is that they only represent an ordinary level of performance. Wheat growers however now talk freely of 10 tonnes per hectare and this is certainly achievable on some fields if not on an overall average year by year. But it is a realistic ambition for the best growers to have firmly in their sights and there is little doubt that, in a very few years, it will be regarded as the attainable norm for the best growers on the best land. How standards have changed! It seems but a very few years ago that we were satisfied with 6 tonnes/ha. Is this relevant to sheep production? Yes, of course it is, for the best wheat growers must be regarded as being in the same class as the best sheep producers. So let me restate the table in terms of the best.

Wheat	9 tonnes/ha	@	£110	=	£990/hectare
Barley	8 tonnes/ha	@	£100	=	£800/hectare
Sugar Beet	50 tonnes/ha	@	£26	=	£1300/hectare
Oilseed Rape	3·8 tonnes/ha	@	£220	=	£836/hectare

You will notice that oilseed rape, which is now a very popular and widespread break crop – far more so than sugar beet which is limited geographically – compares poorly with wheat on the basis of gross sales. And this, of course, takes no account of the cost of growing the two crops. The variable costs of fertilisers and sprays for rape are inevitably very high and question marks must be hanging over the profitability of this crop, particularly as it is so dependent upon EEC subsidy. There looks to be an opportunity for sheep here.

So let us look at comparable figures for sheep on similar land.

Average Performance Figures — MLC Costings for Lowland Spring Lambing Flocks 1987

Overall Stocking rate, ewes per hectare	12·49
Lamb sales: 1·51 @ £38 = £57·38	£716·67/ha
Wool: 12·49 ewes @ £3	£38·22
Ewe Premium: £4·82 per ewe × 12·49	£60·20
	£815·11

At these levels, sheep look a far more attractive alternative to arable crops than they did in 1983 when I last revised these figures. This is especially so with oilseed rape where the price per tonne has been reduced from an average of £270 down to £220 as a result of a cut in the EEC subsidy. However, any sheep farmer with an eye on future prospects should look on the oilseed rape example as a warning of what could happen to sheep subsidies. Sheep are very vulnerable to similar and sudden manipulation in Brussels!

That being said, we must remind ourselves that the figures cited above are for average performance. But we are not interested in being average. What then can we expect from a lowland flock, inwintered and skilfully managed? I think that we must be expecting a stocking rate of 16 ewes per hectare overall on a year round basis, and a lamb sales figure of 1·7. This, I must add, is a performance target in excess of the figures achieved by the top one-third of MLC recorded flocks. And is an indication of the fact that we must aim high. So let us restate the output in terms of these figures:

	MLC top one-third	*High output*
Overall stocking rate, ewes/ha	15·6	16·00
Lamb sales per ewe	1·58	1·7
Lamb sales/ha @ £39·06 per lamb	£962·75	£1062·43
Wool, £3·10 per ewe	£48·36	£49·60
Ewe premium @ £4·85	£75·66	£77·60
	£1086·77	£1189·63

Now we are beginning to get somewhere. On the basis of these figures the sheep, *judged entirely on output*, are competitive with the best of wheat growers, and put barley and oilseed rape in the shade. Only sugar beet does markedly better, but it has to be said that very few sugar beet growers come near an average of 50 tonnes per hectare—far fewer than there are flock masters doing 16 ewes @ 1·7. Furthermore, there are grounds for saying that £110 per tonne of wheat is a pretty optimistic figure looking forward to the next few years. Put wheat at £95 per tonne net after all deductions, worse still £85, and the picture begins to look very different. But then, when one gets into that type of speculation, one wonders what will happen to lamb prices.

Finally, we should recognise that few notable sheep men are achieving results on a large scale that are even better. The ability to achieve, year by year, an output of 1·85 sold per ewe stocked at 16 per hectare is rare, and puts those who do so right at the top of the tree. On the same price basis as before, this gives:

16 ewes per ha @ 1·85 @ £39·06	=	£1156·17
+ 16 fleeces @ £3·10	=	£49·60
+ Ewe Premiums @ £4·85	=	£77·60
		£1283·37

Rare perhaps, but it does show what can be done. But maybe we ought to deal with what we lesser mortals can achieve. From my experience, with reasonable land and the flock inwintered, a target of 15 ewes per hectare at a sale of around 1·6 lambs per ewe tupped is a perfectly reasonable one, not only to aim at, but also to achieve with consistency. A gross return of £1056 a hectare and this achieved with what one might call ordinary lambing percentages. The key is a high stocking rate per hectare.

To emphasise the point – to give the same £1056 per hectare but at 10 ewes per hectare instead of 15 would require a 'lamb sold' figure of 2·5. I *know* that, with reasonable attention to detail, and with good stockmanship, you can achieve 15 ewes per hectare. I would not care to bet too much money on you being able to hit an honest 2·5 year by year, if at all!

THE COMPONENTS OF OUTPUT

I have laid great stress on output as a whole and on the overriding importance of high stocking rates as the essential component of high output. I have also pointed out that lambing percentage does matter. No one can afford to ignore the importance of prolificacy, and as I have pointed out in the chapter on ewe feeding, we are now in a position to think of leaving triplets on the ewe. Moving up from 1·65 to 1·75 at a constant 15 ewes/ha increases gross return by no less than £58 per ha. And moreover, probably achieved at little extra cost. Push it up to 2 from 1·75 at a constant 15 and you add another £146. If wheat growers have a target ambition of 10 tonnes/ha which, on good land is a realiseable ambition, then we sheep producers must be content with nothing less than selling 30 lambs/ha (or 12 lambs per acre).

We must not, however, forget that there are other important components of output to which it is most certainly worth paying attention:

The actual value of each lamb sold. I have taken an average of £39·06 in all these calculations, but there is no reason whatsoever to believe that this will be a constant figure. The factors which will affect what you actually get are:

(1) Quality. We are gradually moving into a period when quality in all its aspects is paid for, and bad quality is penalised. This trend is bound to accelerate.

(2) Whether your lambs are sold finished or as stores. If your feeding system falls down and you are forced to sell as stores, then obviously the return will be less.

(3) Whether you have the opportunity to sell ewe lambs for breeding, when they will usually fetch a premium over the slaughter market.

(4) The time of year when you sell. A high price for early lamb, declining by mid-May and generally not recovering significantly until Christmas, and rising to a peak in March/April.

Then, there is the value of the fleece, which I have taken to be £3·10 per ewe. All right, so this is only £49 out of a total of £1000+ per hectare. But it could so easily be only £2·50 or as much as £4 depending upon the breed of ewe, to say nothing of the care which you devote to its growth and harvesting. A wool cheque for a 700-ewe flock that could be £2,800 is not to be sneezed at.

Finally, there is the sale value of cull ewes. For most lowland flocks this will be at the end of their life, whilst in upland flocks this may well be as draft ewes for further breeding on lower ground. In either case, it is well worth while studying the market so as to be sure of getting the best return.

To give you an example. The majority of lowland flocks are culled, quite properly, after weaning, usually sometime in August. Most people then promptly sell these cull ewes and they have no right to be surprised when they fetch very little money. For one thing the weather is, or ought to be, hot; for another, the schools and lots of other people are on holiday. The main outlet for old ewe mutton is school and factory canteens and preferably in cold weather for stews. August and

September also coincide wtih a flush of lamb marketing, and all in all there could not be a sillier time to sell what is, after all, a low quality product. Keep the cull ewes until January/February and there is a fair chance that they will worth at least 50 per cent more. To do the calculation for you:

- 15 ewes per hectare. 20 per cent annual replacement rate, of which 5 per cent are for deaths, leaves 15 per cent for sale as culls.
- 15 per cent of 15 = 2·25 cull ewes to sell per hectare. If the low value is £18 and the high value is £27, a difference which is perfectly possible, the difference per hectare is between £40–50 and £60–75. Again, not to be sneezed at.

WHY OUTPUT, NOT PROFIT?

I can imagine you saying it! We are, after all, only interested in profit, so why carry on at such length about output. The reason is simple, I don't think that I have ever come across a flock which was losing money because too much was being spent – always the fault lies with not producing enough. This seems to be absolutely fundamental. My experience is well borne out by an analysis of the MLC lowland sheep costings. Looking at the percentage contribution to top one-third superiority, the analysis is as follows:

	Gross Margin per Hectare %
Lambs reared	19
Lamb value	8
Flock replacement cost	13
Feed and forage cost	4
Stocking rate	49
Other factors	7
	100

Gross margin is, of course, a calculation of output less variable costs. The lesson of these figures is that in trying to figure out what it was that put a flock into the 'top one-third' league, success in controlling the variable costs of replacement, food and forage only contributed 17 per cent to the overall picture. No less than 76 per cent was accounted for by superior output.

COSTS

Having said all that, of course costs can be too high. And of course, all of us need to pay scrupulous attention to controlling costs. In particular, I think, to the control of wastage. Feeds are fearsomely expensive yet how often do you see feed wasted, carelessly thrown into troughs so that it spills over, trampled on and lost. The same can be true of drugs and drenches – a general carelessness and often untidyness which costs money either in direct loss or in less efficient usage which shows up in lower output. Most people have a tendency to be untidy, but shepherds often seem to be the worst of the lot!

There is one item of variable cost which is important and has a direct and positive relationship with stocking rate. That is nitrogen fertiliser. The following two figures give the picture precisely.

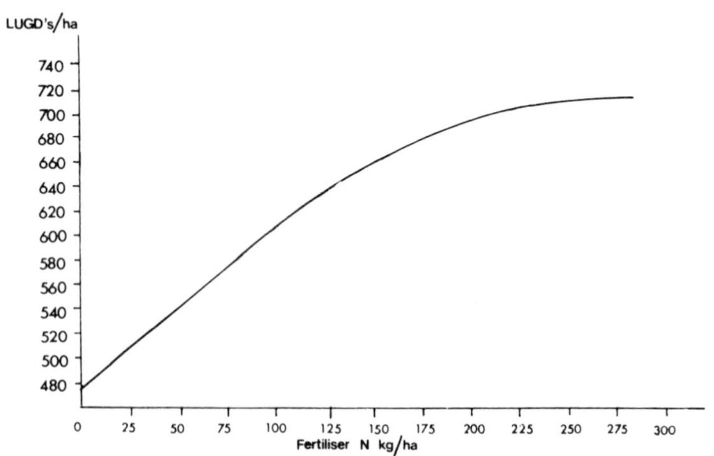

Fig. 7. Average relationship betweeen fertiliser N levels and livestock unit grazing days per hectare.

The lesson is obvious. Do not be mean with nitrogen. Of course the law of diminishing returns comes into play but not before 150–200 kg N per ha.

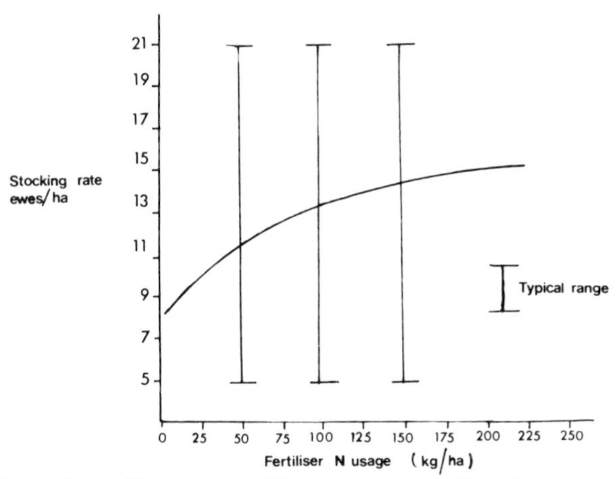

Fig. 8. Range in stocking rates at different levels of fertiliser N usage – Lowland flocks.

Of fixed costs there is little to say except for labour. There is a tendency to employ too much. Good fencing, well-planned handling equipment, sensible work study, the judicious use of casual labour at peak periods – all these together can enable you to cut labour costs back. There really is not any reason why a shepherd should not manage 700 ewes and perhaps 800 depending on the conditions. But make sure that your economics do not lead you to an expensive loss of output.

On land costs, you are either paying far too little if you bought it some long time ago, or far too much if you bought or rented recently. If it is the former one, count your blessings but plan well so that your successors can pay the tax bill! If it is the latter, too bad, you are stuck with it – and you ought to be intensifying your sheep production as hard as you can go.

THE ECONOMIC JUSTIFICATION FOR INWINTERING

Much doubt is cast upon the ability to justify the capital expenditure involved in inwintering. So much so, and because I believe it to be a near essential part of high stocking rates and efficient grassland utilisation, I give below a detailed justification. In doing so, I am taking an extreme example where capital is available to put a brand-new building on a fresh site – the most extravagant situation and therefore the least likely to show any return. The proposal is to provide inwintering for 2,000 ewes on a large chalkland estate where previously most of the ewes had been outwintered. The calculations are as follows:

	Increased Margin/ha
(1) Increased stocking rate from 10 ewes to 15 ewes/ha	
5 ewes extra @ 1·65 @ £38	£313·50
5 ewes fleece @ £3	£15·00
(2) Increased lambing percentage due to more lambs sȧved + 5%	
15 ewes/ha @ 5% @ £38	£28·50
(3) Elimination of bad weather risk. The farm is exposed and is very vulnerable to a really bad winter and spring Assuming one bad winter in six Loss of 10% lambing	
15 ewes/ha × 10% @ £38 ÷ 6	£9·50
(4) Saving in food during winter due to avoidance of waste, say £0·75 per ewe × 15	£11·25
(5) Outwintering necessitates 'sacrifice areas' which then have to be drilled with spring barley instead of winter wheat; 60 ha would be needed	

Wheat

Output: 7 tonnes @ £110	=	£770·00
Less variable costs		£190·00
Gross Margin	=	£580·00

Spring Barley

Output: 5.5 tonnes @ £120	=	£660·00
Less variable costs		£160·00
Gross Margin	=	£500·00

Economic disadvantage of growing Spring Barley rather than Winter Wheat:

£580 − £500 = £80/ha × 60 ha = £4800 ÷ 2000 ewes = £2·40 per ewe × 15 ewes/ha	£36·00

(6) Longer average life of ewes – lower depreciation due to ability to keep in a favoured pen of old ewes which would otherwise have been culled; say a reduced culling rate would apply to only half the flock	
If ewe depreciation is £75–£30 = £45	
Over 4 years = £11·25	
Over 5 years = £9·00	
15 ewes/ha × £2·25 × half the flock	£16·87
Therefore total savings or increase in output	£430·62
per ewe @ 15 ewes/ha	£28·70

Other Advantages and Savings

There are other advantages and savings which are difficult to quantify, but are nonetheless very real:

(a) Better utilisation of labour. The flock is centralised, rather than scattered. Less time needed for feeding. More time available for attention to detail.

(b) Saving in fuel. Shepherd does not have to travel by tractor to see to his sheep.

(c) Better working conditions for the shepherd.

(d) Higher grass output in the critical months of April and May leading to better milk yield and therefore faster growth in the lambs.

(e) Other use for the building during the summer, e.g., machinery or fertiliser storage.

VARIABLE COST INCREASES

There have to be some increases due to the change in management to set against the increase in output:

(a) 3 weeks' extra feed to ewes @ ½ kg head day to support higher lambing percentage	£1·68/ewe
(b) Additional food for finishing extra lambs	£0·75/ewe
(c) Increase in vet and medicines allowance	£1·50/ewe
(d) Increased cost of N. fertiliser @ £15 hectare	£1·33/ewe
(e) Sundry costs	£1·00/ewe
	£6·26/ewe

Summary

Savings and increased output	£28·70/ewe
Less increased variable costs	£6·26/ewe
Increased margin	£22·44/ewe

Annual depreciation and interest on the building costing £50 per ewe housed, or £100,000 – amortised over 15 years @ 14 per cent interest is £16,300. For 2,000 ewes this represents an annual charge of £8·15 per ewe. Therefore the increased margin after depreciation and interest is £22·44 – £8·15 = £14·29 per ewe. On 2,000 ewes, this gives a net increase in the flock margin of £28,580.

I emphasise that this is an extreme example. It is still possible, even at 1985 prices, to house ewes in perfectly adequate buildings for a capital cost of a good deal less than £50 per ewe. But if that level of expenditure is justifiable, then how much more so for something rather less lavish. The case is I think quite clear but, and it is a fact that must never be forgotten, the increased return depends upon increasing the output as a result. Go to the expense of inwintering, and do nothing else and all that will have happened will be an increase in costs. Inwintering is not just an easier winter – it opens the door to a completely different style of management.

CAPITAL COSTS OF A SHEEP UNIT

Finally, a word on capital. I suppose one really comes up against reality when asked, how can you justify an outlay of such size if starting from scratch with a new enterprise.

Fifteen ewes to the hectare – that on its own probably represents a value of £50 per head taken across a range of ages from ewe lamb replacements to, say, six-year-old ewes: £750 per hectare for the sheep alone. And if they were all bought today in 1983 as gimmers, they would probably cost £85 and that would be £1,275. Add to that a modest £25 per ewe for inwintering, which is another £375. Then there is fencing, handling pens, and other sundry equipment including a tractor for the shepherd, say another £15,000 and that would assume using the second-hand market as much as possible.

So for a 700 ewe unit employing one shepherd and using 47 hectares of the farm, the sums look like this:

700 ewes @ average £50	£35,000
20 rams @ average £250	£5,000
Inwintering shed and equipment @ £25 per ewe	£17,500
Fencing, handling, tractor etc.	£15,000
	£72,500
or	£1,542 per hectare.

And these, remember, are minimum figures – it would be quite easy to spend much more. I have to say that I find these figures quite horrifying. I would find it very difficult indeed to prepare any sort of budget based upon this capital expenditure which made any sense. The conclusion would be that on today's returns, the investment could not be recommended.

Yet if I look closely at any proposition to set up a dairy herd for example – even if that were possible under current quota arrangements – or a pig unit of 300 sows, or with corn growing, the investment in corn handling, drying and storage, I know that I would come to the same conclusion.

This says something about the financial position of farming in Britain which does not make much economic sense, that the industry as a whole, and we individuals within it, are only profitable because we are operating on historic costs. My sheep unit is 'profitable' only because it has been in existence for a long time. Many of my ewes cost me not £50 but between £25 and £30 as ewe lambs five years ago, and will be sold off as culls for as much as I paid for them. My inwintering sheds, which were new at the time, cost me less than £10 per head housed. And so I could go on. My investment was high at the time I made it, but inflation has made it look very attractive. The truth is that a great deal of our agricultural investment has been built upon inflation and capital allowances against taxation. Furthermore, for very many years we were in the attractive position whereby inflation comfortably exceeded interest rates, i.e. we profited from 'negative' interest charges. The position today is very different indeed. With inflation strictly under control at around 5 per cent and interest rates consistently in double figures, investment on borrowed money has to be looked at with a very jaundiced eye.

These are gloomy thoughts and they make farming look a totally different world from the one I entered first as a student in 1945 and as a farmer in my own right in 1959. But we have to live and operate in today's world. If no investment in farming makes any sense, then is the conclusion that you should up sticks and be off elsewhere? No, because I suspect that exactly the same position is to be found in industry. Apply inflation accounting and you would bankrupt British industry at a stroke!

There are however other reasons for discounting such total pessimism. Of course it is true that British farming as a whole faces a future that looks a good deal less promising than any period since the war. There is the shadow of surplus production hanging over us together with the determination of Government to impose strict financial discipline upon us. But that is not to say that there will be no opportunities – rather the reverse, because it is often true that the prospects for the enterprising are greater in times of adversity than in the fat years. Sheep, as an enterprise, are susceptible to the application of hard work and the do-it-yourself economy. The purchase of draft ewes at low prices; the adaption of existing buildings – all these can change the budget significantly and enable the enthusiast to build up his enterprise. And at least you can say this for sheep – your capital, or most of it, will be walking about on four legs. It is therefore realisable, unlike a big investment in corn drying and storage. Lastly, and by no means least, you are in on an enterprise where the future looks more secure than with some. At least we are in deficit production in Europe and there are still imports to replace and export markets to exploit. But only if we get our product right – and that means *lean*.

Chapter 17

THE FRENCH SHEEP INDUSTRY

ENTRY INTO the Common Market has brought about some dramatic changes in British Agriculture as a whole. Britain's traditional exploitation of the cheap food market of the world has had to be adapted to the basic philosophy of Europe. That is to say, a liberalisation of trade within the Ten (soon to be Twelve), the prosperity of the whole being protected by controls on imports from outside.

British farmers saw our entry into the EEC as an opportunity to exploit their own efficiency in a far wider market, and moreover a market which was accustomed to paying high prices for food. In no section of the British agricultural industry was this more true than amongst sheep producers. In particular, they looked at France, after Britain the biggest market for lamb in Europe. The original, and wildly optimistic, expectations of a bonanza were never realised. It was naive to think that they ever would have been. Nevertheless the French sheep market has had an increasingly dominant influence on the developing prosperity of British sheep farmers, and it behoves us to know something in detail of how the French sheep 'mind' ticks.

SOME STATISTICS

Numerically, there are near enough half the number of sheep in France that there are in the UK, that is to say, 11 million head of all ages and types at January 1st. Of this global figure, 6·8 million are breeding ewes; and in this figure are included 0·8 million ewes which are milked for cheese production.

The annual production of mutton and lamb is of the order of 150,000 tonnes and this supplies approximately two-thirds of a domestic consumption of 3·9 kg/head/annum. 50,000 tonnes of sheepmeat are imported and another 8,000 tonnes of carcase equivalent come in as store lambs.

At first sight, it would appear that France, with approximately the same human population as the UK, has a sheep industry about half as big and half as important. Since 1974, sheep numbers have been rising but only very slowly, and with the present mood of pessimism amongst producers I would not think that any significant increase will be seen in the near future. Consumption, on the other hand, is rising and quite steeply at that, and furthermore, it is rising against the background of high prices relative to other meats.

The French sheep industry is most definitely *not,* however, a mirror image of its British counterpart on half the scale.

THE STRUCTURE

Here more than anywhere France differs from Britain. It is a difference which is quite fundamental and it affects the whole efficiency of their sheep production.

We, in Britain, have, over the years and particularly since 1947 followed a policy of supporting hill farming along the lines of making best use of the natural resources to be found in the hills, that is to say, the land and the hardy pure breeds of sheep that are indigenous to each area. Farm size has been allowed to find its own level, responding to economic pressures, and the result is a reasonably healthy and viable hill farming business. It is this which forms the essential base of the triangle on which our sheep industry is built. For the draft hill ewe, put to the 'long-wool' crossing ram, gives us the productive half-bred ewe, which in turn is the whole basis of our lowland lamb production.

There is nothing comparable to this structure to be found in France. Hill farming land there certainly is, and much of it of inherently good quality which could support large numbers of sheep such as the Scottish Blackface. Why, then, is it not used in the same way as our hill land? The answer lies partly in the French law of inheritance which tends to lead to fragmentation of holdings, and partly in French government policy in response to fragmentation.

French hill farms are small and peasant in character, similar if you like, to Scottish crofts. But whereas history dealt most cruelly with the crofting problem in the Scottish Highlands, the French variety remained. Successive French governments have had to face a severe social problem of low incomes and an ageing population in the difficult regions, particularly those which are away from the tourist ski resorts.

A typical example is the Millevaches plateau lying astride the boundary dividing the Departements of the Creuze and Correze in the eastern Limousin. At an altitude of 600–1,000 m and a rainfall of 1,500 mm it is an area growing grass and heather, and with plenty of trees, it strongly resembles parts of Wales and Scotland. With good management, proper fencing and the right breeds of sheep and, above all, big farms, it could be a very attractive hill farming area. Instead it is very depressed, and more and more of it is being abandoned, sometimes to professional forestry and sometimes just left to fall down to scrub. The problem can only get worse as the young people leave for the towns.

Now a social problem on the Millevaches is nothing new. Traditionally, it is an area that has always exported its people. For some curious reason it has had the reputation over the centuries of providing the masons doing building work in Paris. But until relatively recently, this export was solidly based on an indigenous population on the peasant farms of these hills. The modern industry of food production has done no good at all to the way of life which areas like these have provided.

French governments have, of course, been highly aware of the problems caused by an impoverished agriculture, and it is true to say that they have tackled them with considerable vigour and no little financial help. But if an outsider may be permitted to say so, they have made a fundamental and fatal mistake.

For instead of recognising that the area could not support the existing number of small farms and facing up to the consequences, they attempted to maintain the existing numbers of farms by intensifying the individual output on each farm. I saw the result of this policy some years ago when I was shown, with great pride, what was held up as a model farm. At 120 ha it was a lot bigger than many, but the land was not used. Instead a battery veal calf unit was installed in the old buildings; all the calves and all the feeds were imported up hill. A nonsense on a hill farm. Then I was shown the sheep shed and here at last I thought I shall see something relevant to the farm. Not a bit of it! A palace of a building, brand-new and costing a fortune, it contained lowland ewes housed most of the year and lambing out of season.

It was all really rather sad. For such production could have been carried out so much more economically in the lowlands where the maize is grown and the climate so much better.

But the really damaging result of this policy, which is grant-aided at a high level, is that the French breeds of hill sheep have largely disappeared. For there is no point in using a hardy but relatively low productive ewe for out of season production in a shed – even if that shed is 1000 m up. The result is that they have been crossed and crossed again, so that they are becoming lowland sheep.

BREEDS OF SHEEP

I have devoted a lot of space to French hill farming because the policies that have been followed have deprived their lowland lamb producers of any decent sheep with which to do the job. Just imagine what our lowland sheep industry would look like if we had not got the half-bred ewe!

So the lowland sheepman has been left, as it were, to his own devices, to find the breed most suited to his conditions. These have included, amongst other things, a market which has paid extravagant attention to good conformation and high carcase quality and has been prepared to pay for it. The result is that the basis for the commercial lowland flock has been a small 'Down' type of ewe of good conformation. Most producers have bred their own replacements and have followed a system of using a different breed of fat lamb sire each year and keeping his ewe lambs. The result it an unimaginable cocktail made up of constant mixing of about half-a-dozen meat breeds, and a complete absence of hybrid vigour.

Now France has got some extremely good meat breeds such as the Ile de France, the Berrichon du Cher, the Charollais, the Vendean and now the Suffolk – but quite rightly they have been bred for their carcase qualities. They have not been selected for their prolificacy or milk yield, even if it were possible to combine these qualities with extremes of carcase quality.

So the French commercial lamb producer is straight away at a very real disadvantage. His commercial ewes just are not productive in the way that our half-bred ewes are, and he has no source of supply to turn to where he could get something better. Furthermore, until the last year or so, he has not really had to bother. Lamb prices and profitability were high enough to compensate for low output – but that is changing fast, and French lamb production today at their level of cost is a very marginal enterprise.

French research has been well aware of the situation but, like some of their opposite numbers this side of the Channel, has sought the dramatic solution. They have chosen to work with the Romanov rather them the Finnish Landrace but the result has been the same – a sheep not suited to the abilities or to the requirements of the great bulk of producers.

THE ECONOMIC ENVIRONMENT

The French sheep man has to work and hopefully prosper under the economic conditions in which he finds himself. Only to a limited extent can he alter them. Amongst those conditions there is one which is a great constraint – that is, the restriction which effectively prevents farm size rising above a certain, and rather low, level. This is not the place to go into the detail but, basically, farm size cannot respond to economic pressure and the economics of scale in the way which they have done here. So there is the limitation of farm size.

On the other hand, the French sheep political organisation, the FNO, has taken full advantage of two important assets, for them, in the political and economic climate of France. The first is that lamb is regarded as a luxury meat and is readily saleable at prices which would put it out of the market in this country. Why, they ask with Gallic logic, kill the golden goose by importing some cheap eggs?

The second is that livestock production generally, and beef and sheep in particular, employs a politically important section of the population. Their votes are needed to balance the power of the town voter.

There is an important element in French political thinking which has no equal in Britain. It is that, for political stability – in a word, freedom from revolution – it is highly prudent to keep a significant slice of the population in the countryside. Many of the rural areas which are important in this context are sheep areas. The FNO has played this card with great skill. It would have done us no harm at all to have had agri-politicians of similar ability.

So the economic and political environment has tended to keep sheep farms small and politically important. It has also insulated the French sheep producer from evolution and ill prepared for the rougher times which lie ahead.

THE FRENCH MARKET

The French producer has been able to say with complete justification that he had a special market: a demand for really high quality, at high prices and which justified expensive and out-of-season production. It was most definitely not a market for the frozen lamb nor for supermarket selling.

Whilst all that remains true to a degree, changes are coming fast. The small butchers shop is coming under pressure. Even in the small villages, housewives get into their cars once a week to go to town to shop in the hypermarket. Increasingly, you find a deep freeze in French homes. The French housewife – sadly perhaps – is started along the road long since taken by her British cousins. The writing is truly on the wall.

THE DILEMMA

First of all, there's a dilemma for the French Government. How to keep the difficult areas electorally happy in a political climate which is very finely balanced between left and right. How to bring about progress in grassland production without a damaging acceleration in rural depopulation. Does it make any sense to take people out of these areas and put them into an urban dole queue?

Then, there's a dilemma for the sheep producers' organisation. How to resist the pressures and the temptations to use their real political muscle to maintain, in the short term, their defensive position. Rather than concentrate on helping their fellows to produce competitively, which is essentially a longer term and much less dramatic task, but eventually much more rewarding.

And all this against the background of the real threat coming from Britain. This frankly is what sticks in their gullets. For, to their eyes, here is Britain the most Un-Community-minded of the Ten, whose attitude in the past has been aggressive and wholly partisan, and who imports 30 per cent of her lamb on the cheap from New Zealand, but expects to export 20 per cent and more to exploit the French market. This is something they feel that they can quite legitimately resist.

THE FUTURE

Any thinking French sheep politician realises that change has to come. Not only from the EEC, and that must now be certain after the ruling on free movement of potatoes, but also from the growing influence of French consumerism. But at what speed and when?

Let it be said straight away that French methods are changing fast. A younger generation is shedding some of the traditional inhibitions. The advisory services are in the main staffed with young men of enthusiasm, though they have nothing to compare with the efficiency that we accept almost carelessly from ADAS and MLC.

Also, there is nothing like a bit of economic squeeze to increase efficiency. And this is just what is happening. French grassland management is, with some honourable exceptions, of a deplorably low standard. In complete contrast to the super efficiency of their arable farming, which is second to none in Europe.

We in Britain have a tendency to comfort ourselves with this fact. We are, we claim, amongst the most successful grassland managers in the world. Dangerous complacency! French grassland productivity is so low, particularly in the sheep areas, that it could be doubled with ease and without the application of any particular skill. Sixty kilos of N instead of none, plus the use of a grass topper would work wonders. This illustrates the potential that lies within French grassland agriculture, and it should be a salutory thought for we British who see a rightful place for ourselves in the French market. At present they supply 70 per cent of their own consumption. It would not be difficult to see that going up to 100 per cent in a relatively short space of time, and even becoming net exporters in due course.

We are very competitive at the present due to the quality of our sheep *and* to the relatively low costs which we have enjoyed. Now our costs are shooting up and must before long reach French levels. How would you like to pay 38 per cent on top of your mens' salaries, as Social Security payments?

It is not impossible that we may find by the time the French market is fully and freely open to us, that it is nowhere near as attractive as it appears at the moment. That is not to say that we should not work harder still to build on the bridgehead already established by the successful few of our meat exporters. Of course, we should. And the rewards will be there not only for carcase lamb but eventually – when the barriers of tradition have been knocked down in Paris – to exploit to the full the really big opportunity which is to provide them with the quality commercial breeding ewes, and store lambs out of our hills.

Chapter 18

THE FUTURE

THE POLITICAL BACKGROUND

ANY CONSIDERATION of the way our sheep industry is going to develop must start with the recognition that we, as farmers in general and sheep producers in particular, are only partly masters of our own destiny. We have to make guesses about the way our future governments are going to shape agricultural policies in response to pressures which are almost entirely non-agricultural. And, in a wider context, to speculate on the evolution of the Common Market as well as developing opportunities in the Middle East. It is not simply a question of whether we shall be allowed to compete on equal terms with the French. We must admit with honesty that our future depends very much on the way governments are prepared to support us. For support we must definitely get, whether it be in the form of actual financial subsidy or as a longer term guarantee of fall-back prices.

We farmers have no cause to complain about this, nor indeed have we any reason to be ashamed. By contrast the virtual abandonment of British agriculture between the two wars had nothing to commend it whatsoever. The dangers into which that policy led us as a nation, when we had so destroyed our food production that we very nearly lost the war due to starvation, were fully recognised in the immediate post-war years. That recognition was enshrined in the 1947 Agriculture Act and today, 38 years later, no politician or economist would dispute the basic premise that agriculture merits support. It is a two-way bargain which gives real benefit to Government coffers and to consumers pocket alike as well as the farmer's personal prosperity. The argument is about detail, not about principle.

So what about the detail. Now after so many years, we can be as sure as anyone ever can be that Britain will remain in the

Common Market. And more than that, we are likely to become more involved as constructive participators in every way and that must include the Common Agricultural Policy. Those who argue that the CAP is unworkable and is bound to destroy itself by drowning in surpluses miss the essential point. European agriculture is about much more than supporting peasants on tiny and uneconomic parcels of land, even if that were true. It is about National and Community security of food supplies from within. The reduction to the lowest possible level of reliance on imports from abroad, which not only cost money but depend upon sea transport vulnerable to attack or threat of attack. To say nothing of a vulnerability of fuel prices and supply which make the proposition that we should buy our food supplies in the low-cost areas of Australasia and South America look imprudent, to say the least.

The CAP is also about political stability. Not in the sense of keeping governments, or even Ministers of Agriculture, in power. It is the recognition of the dangers of concentrating too high a proportion of the population in the inhuman environment of our modern cities. Dangers which are far greater where unemployment is at a high level. It makes no sense whatsoever to squeeze people out of agriculture merely to stand them on a street corner in an industrial town where there is no work for them. Quite apart from the moral destruction which this brings about, it probably costs as much to support them via unemployment and other social benefits as it does on their farms.

So the CAP is about far more than food prices and surpluses. It is very much about the political and social environment in which we live. Because of that, solutions will be found to its present problems. What those solutions will be is, mercifully, not the subject of this book It does, however, lead me to be optimistic that European agriculture has a viable and prosperous future. And that, despite the rising power of the consumer voice at Brussels, there will be no return to the 'Cheap Food Policy' which dominated British agricultural economics right up until the mid-1970s.

ADVANTAGES TO SHEEP PRODUCERS

As sheep producers, rather than as farmers in general, we

can face the future in the knowledge that we have some very considerable advantages on our side. Our main product, lamb and mutton, is virtually the only temperate foodstuff which is in deficit supply in the Community, only about 70 per cent of present consumption. True, that deficit could very quickly be made up if even average performance in Britain were to be achieved elsewhere. On the other hand, provided that price relative to the other meats can be kept reasonably competitive, there is every reason to believe that consumption could be increased despite a tendency for it to fall in Britain. Particularly is this so in Germany, but such an increase would demand a real effort in promotion. But then why not promote? So at least, for the moment, we are not bugged by the surplus worry.

Then our secondary product, wool, has stood up surprisingly well in the face of competition from artificial fibres. There was a time when it was thought that wool would be almost entirely superseded, but its natural qualities have kept it a real place in the market. I see no reason why this should not continue to be the case always provided we look after it and market it properly. In Britain we have a real advantage in our Wool Marketing Board. A copy-book piece of orderly and intelligent marketing if ever there was one, which adds greatly to the value of British wool. It is interesting to compare French and British returns. Whilst during the late 70s French lamb prices have been consistently 50 per cent higher than those in Britain, wool has only been worth half in France what it would have fetched here. Our wool marketing system ought to be copied elsewhere in Europe, and under no circumstances should we ever allow any bureaucrat to talk us out of this Board.

So, politically, we sheep producers are not a cause for worry or comment as in the case with milk. No one in London or Brussels is scheming to reduce or control us, and let us be thankful for that! Successive British governments have realised that sheep production could be expanded with justification. Both *Food from our Own Resources* and its successor, in February 1979, *Farming and the Nation,* projected significant increases in the national flock. These increases have now come about. Have we reached a ceiling. I wonder? Certainly not in so far as the technical potential is concerned. And we have the gap left by a reduction in the national dairy herd, and perhaps also before too long, a significant cutback in the cereal acreage. But room for extra sheep on the ground is one thing: a profit-earning place for them in the artificially supported market is very much another. Much will depend upon our exploitation of market trends and changes in public taste. Can we be optimistic about that? I would so very much like to be, for we have basically what is such a very good product. Not for nothing do the French refer to lamb as 'La Viande Royale' – the meat fit for Kings. In its different forms, properly cooked and presented, there is nothing more succulent and tasty. The fact that we see consumption falling in Britain is due far more to our own incompetence as an industry than to faults in lamb as a meat.

OUR PLACE IN WORLD MARKETS

What is going to happen to New Zealand lamb? Is it going to continue to meet some 30 per cent of our home market consumption? The answer is most certainly yes, if the New Zealanders have got anything to do with it. Some items of news from that far-off country of low cost production ought to be causing us no little concern. They have completely restructured their system of carcase grading so as to improve quality by eliminating fat. New technology has been introduced into their meat plants which will result in up to half their exports to the United Kingdom coming broken down into primal cuts. Finally, and perhaps worst of all, there is the contract won by New Zealand to supply boned-out lamb to Matthews Norfolk Farms – the business which has pioneered the marketing of turkey joints so successfully. Worst of all I say, not just because it is New Zealand lamb but because Matthews *wanted to buy British*. But not one British EEC Approved abattoir was prepared to give him a quotation.

With news like that, we deserve to see the New Zealanders taking more of our market. One cannot have other than tremendous respect for the tenacity and aggression of New Zealand marketing. It is at least possible that they will eventually get a substantial foothold with competitively priced frozen lamb joints and deboned steaks on the supermarket shelves of Europe as tastes change. And if they do that, with their great disadvantage of huge distance from their markets, we shall have no one to blame but ourselves.

Europe, however, is far from being the world. Increasing wealth in the Middle East, where mutton is the first-choice meat, is leading to greatly increased demand. New Zealand currently exports 31 million lambs and 7 million sheep annually, of which 53 per cent is destined for the EEC but great efforts are being made to develop trade in Iran, Saudi Arabia, Japan and even Russia. Indeed New Zealand is deliberately following a policy of diversifying away from the British market.

Frankly, any estimation of the future trends in world markets is all guesswork. The past is no guide to us in Britain for the simple reason that our sheep industry was entirely inward looking. Certainly that is no longer the case. Of course, the home market will remain our most important outlet; a market which we should seek to cultivate and indeed one where we should be looking to replace a significant proportion of present imports. But it will be a base on which we should be working for an expansion into export. My guess is that it will be the export element of our marketing which will put the real push into any flock expansion that will take place.

THE UNTAPPED POTENTIAL

Looking back over farming since the end of the War, the greatest impression that anyone must have is one of tremendous change in technique and management. We are just not in the same world as we were in during the 50s. Developments in mechanisation, plant breeding, control of disease, nutrition – all these have brought about change that we would have thought impossible twenty or even ten years ago. And the speed of this change shows no sign of slackening. This has been and continues to be true of cereal production, of dairy husbandry, and of pigs. In almost every sector, yield has consistently risen, real costs have fallen and the whole management of any one enterprise has become precise and highly demanding both of skills and capital.

In a curious way, sheep have remained, if not completely outside of this change and development, at least in a very real way to one side of it. Unlike the picture with increasing milk yields per cow, more pigs per sow per year, the national flock productivity has hardly changed in twenty years. Is this because the sheep is essentially an animal unsuited to the exploitation of modern technology; or is it because few people have thought it worthwhile to put the effort into it?

Anyone who has got this far in this book will, I hope have gained the impression that I firmly believe that a significant upsurge in our national sheep production is both possible and desirable. Furthermore, that it can come about without the need for any dramatic breakthrough in technology. What is required is an abandonment of the conservative pessimism which has, perhaps with some good reason, permeated the industry during the past three decades.

To quote a Nuffield Scholar, my eldest son, recently returned from a tour of New Zealand and Australia to study lamb marketing: 'Our sheep industry has the expertise, the genetic material, and the technology to make lamb a highly attractive and competitive meat. We must try and pull ourselves out of the stifling traditionalism that is letting the white meat industries, with their aggressive marketing and versatility of product, and the health movements, whittle away at our livelihoods.'

So the know-how to achieve an increase in both the quantity and quality of production exists. Let me try and summarise from where and how this might come about.

(1) *Flock Productivity*

Taking very round figures so as to argue the case simply, the national (UK) situation is as follows. We have 15.5 million ewes and we slaughter annually, approximately 13.9 million 'other sheep and lambs'. This figure includes hoggetts. In addition, ewe lambs to the tune of around 20 per cent of the total ewe flock will be retained as replacements – say 3.1 million. Add the 13.9 and the 3.1 together to give an overall production of 17.0 million lambs per year, which gives a true lambing percentage (lambs sold to total ewe flock) of only 109. Hardly an earth-shattering performance. Allowing for the fact that half the national flock is to be found in the hills where productivity is unlikely to be more than 85 per cent, then straight arithmetic shows that the 7.5 million ewes in the lowlands achieve an annual productivity of no more than 134 per cent.

We know from MLC costed lowland flocks in 1983 that production on average was 149 per cent. The gap between the national average and this figure represents a tremendous potential of lamb production that is not being realised. And no one surely would claim that 149 per cent was in any way an exceptional level of performance. There are plenty of competent flockmasters who are doing significantly better, and that without the application of anything out of the ordinary in the way of technique or capital employed.

If our ambition were modest indeed and were to be no more than to raise national lowland productivity from 134 per cent to 149 per cent, then we would increase output by no less than 1.12 million lambs a year. At an average of 20 kg per carcase, that represents an extra 22,400 tonnes of lamb meat annually.

This is only in the lowlands. There is no need to be pessimistic about the output potential from the hills. Of course, it will always be limited by the severity of the environment. Nevertheless, the gap between average performance and the reasonably good is, I suspect still large. MLC-costed flocks may well give a less reliable indication of the position than is the case in the lowlands simply because the really poor hill conditions are probably not included.

Even with this qualification, the difference is significant: 93 per cent of lambs reared for the average of all costed flocks as against 105 per cent for the top one-third. Surely it would not be unrealistic to hope for an increase from the overall average of perhaps 80 per cent up to 90 per cent. That amounts to an increase of no less than 700,000 lambs annually.

Taking lowland, upland and hill together, a modest estimate of potential increased production resulting from upgrading standards of performance to levels which are perfectly easily attainable is of the order of an extra 2 million lambs annually. It must surely be the main aim of the Government, Advisory Services and the industry itself to bring this about.

(2) *Increased Ewe Numbers*

If profitability increases due to the increased productivity, we could expect to see an increase in the national flock. Furthermore, as both the milk and cereal sectors come under increasing pressure, there will be an incentive to increase sheep at the expense of either or both.

Stocking rates too, like individual ewe productivity, still have a long way to go. Again using MLC costings, stocking rates of lowland flocks have increased on average from 10.4 ewes/ha in 1977 to 13.1 ewes/ha in 1983. This figure could surely go to the 16.6 ewes/ha achieved by the top one-third flocks. An increase which, at 1.49 lambs sold per hectare. Something like an extra £200 per hectare in gross income!

(3) *Better Grassland Management*

We pride ourselves on being pretty good grass managers, and by comparison with the standards that are all too evident in France, for example, that is no doubt true. But again the gap between the average and the good is significant. One has only to drive about the British countryside to be aware of the large areas of pasture that are far from being what they should be. I get the impression that the bulk of our well-managed grass is dairy grass. Perhaps what we need is a milkman's outlook.

(4) *The Under-utilised Hills*

As a national resource, our hills and uplands must be without parallel. A resource of great richness only a small part of which is utilised. The knowledge is there that could add greatly to both sheep and beef cattle production. Fencing, planting of shelter belts, strategic spot improvement of pasture, phosphate and lime, the use of pioneer crops for lamb finishing, perhaps inwintering – all these are well known and the results documentated. The barrier is always that improvement on the hill creates a cash-flow famine and it makes more economic sense to keep expenditure to a minimum rather than attempt to push output up. The fact remains that it could be done if ever Government choose to provide the means.

(5) *Unused Food in the Arable Areas*

I drive around Lincolnshire with one eye firmly fixed on my neighbours' crops, comparing and criticising and usually regretting that I am not doing better. With my other eye. I am looking at the opportunities that exist for fattening lambs. Time was when Lincolnshire, Yorkshire and East Anglia were the home of great folded flocks, an essential complement to the arable crops. Those days have gone and no doubt will not return. But it is still true that there is room for a considerable number of sheep on food opportunities that are otherwise wasted. Sugar-beet tops ploughed in and not fed off. Early crops of peas, potatoes and winter barley harvested in ample time for a catch crop of stubble turnips to be sown and before the land is required for the following cash crop. The technique of catch cropping is a science which is much neglected and yet which has much to contribute both in profit and to soil fertility.

WILL THE POTENTIAL BE TAPPED?

That the physical potential is great, there can be little doubt. There is also little doubt in my mind that the conditions are favourable – mainly due to the opening up of export markets – for a move towards a gradual realisation of some small part of the potential. It will come about, not because of any conscious act of Government, but because of better returns from a wider market. I am tempted to say that, hopefully, these wider markets will be skilfully managed and exploited. That in the process of stimulation, too much extra lamb will not be produced so as to oversupply a market which fails to keep pace with production. It may seem over-pessimistic to sound such a note of caution when we only produce some 70 per cent of our own national consumption. Yet the balance is delicate and we are in no position really to do any management of the market at all. We must rely on the skill and entrepreneurial enterprise of our main exporting abbatoir/wholesalers on whom we depend for opening up these extra markets.

What we as sheep producers can do is to put our weight behind extensive promotion of British lamb in overseas markets. I must admit to being in some grave doubt as to the justification for meat promotion on the home market where often enough this merely persuades a housewife to change from one kind of meat to another. But I am in no doubt at all about the validity of promotion in overseas markets where we are seeking not only to increase consumption, but actually to take the market from someone else's product.

Perhaps a much more fundamental question is whether a British government will ever consider it worth while to do much more than allow the industry to take the opportunities that come to it. For to unlock the potential in any real way would involve government aid on a generous scale, principally to overcome the cash-flow problems which are the direct consequence of improvement in the hills. Could Britain carry 20 million ewes without seriously reducing other forms of production? I think she could but it would require a transformation in government policy which we are hardly likely to see. What is much more probable is that we shall see a slow climb upwards giving us perhaps a 10 per cent increase in ewe numbers by the late 80s.

WILL TECHNIQUES CHANGE?

All through this book I have argued that the need for dramatic change is far from proven. That good husbandry and stockmanship, linked to a determination to apply the same levels of attention to detail that are automatically given to the dairy cow, are all that are needed to transform the sheep flock into a competitive profit earner. Nevertheless, there are some changes in the pipeline to which at least I should draw your attention.

(1) *Winter Shearing*

This is a recent development which was pioneered by the Ministry of Agriculture Experimental Husbandry Farm at Drayton, near Stratford on Avon. We took it up in 1979 for the first time and quickly became enthusiastic about the advantages. It is of course only relevant for ewes which are winter housed. Our ewes are shorn in January prior to the lambing in late March. This mid pregnancy timing is important, for it is at this point that there is no risk to unborn lambs. All our ewes have been winter shorn now for five years and we shall never go back to summer shearing.

Briefly the advantages are:

(a) The ewes are much easier 'to see' in the shed. I emphasise 'to see' because suddenly one becomes aware of how difficult it was previously to see exactly what was happening to any one individual. Carrying a heavy fleece, a ewe can lose a great deal of condition before it becomes obvious. Now shorn, such a ewe is noticed immediately and is dealt with. This advantage carries right through to lambing.

(b) Respiration rate and skin temperature are both lower. The difference is very noticeable, an unshorn ewe appears to have almost laboured breathing alongside a shorn ewe and this even in a well ventilated shed.

(c) Food consumption is higher, and whilst this can hardly be considered an advantage in itself, there is no doubt that this is one of the main reasons why lambs from shorn ewes are born with a significantly higher birthweight. The increased consumption during the housed period may be as high as 11 per cent.

(d) More lambs are born alive at a higher weight and survival to weaning is significantly better. This must be partly due to the higher birthweight, but is also due I am sure to the fact that the lambs can much more easily suckle their mothers.

(e) I get the impression that the ewes milk better and I think that this also must be due to higher food intakes and a more efficient utilisation of that food.

(f) Shorn ewes require less space in the shed and less length of trough.

These are real advantages. Are there any snags?

Only in the first year, when the fleece is no more than 7 months old is there any reduction in wool yield. The other fear that the shorn ewes would suffer from the cold certainly has not been realised. The winter, January to March 1979, was as severe as any is likely to be, and the weather following turning out was as bad or worse as I have ever known it. But we had no problems and certainly no increased levels of mastitis. My conclusion is that we shall see a great deal more of January shearing where ewes are housed. And as we are unquestionably going to see a real increase in winter housing, the two will go together.

(2) *A Change to heavy Carcases*

A change is taking place in the big factory abattoirs which is likely to have far reaching repercussions right back to the breeding and selection of the final crossing ram. More and more the work of actual butchering is taking place in the meat preparation plant adjacent to the slaughterhouse. Instead of whole or half carcases going out to the butchers' shops, cutting down is done into at least primal joints before despatch. Butchers' shops are becoming mere supermarket shelves retailing ready packed joints. This move has become inevitable as skilled butchers have become both rare and expensive. Sadly the old-fashioned butcher's shop, where the skilled butcher/owner discussed requirements for the Sunday joint with his housewife customer, is giving way to 'progress'. How sad, but we had better recognise that this is happening. The eventual outcome of this is that the carcase should not only be jointed on the factory line but should also be boned and packed. That way the retail shop is provided with the packs that it requires, the transport costs are kept to a minimum as unusable fat and bone is left behind, and finally the abattoir makes full use of these by-products.

The effect of this change to we producers could be significant, for the most economical carcase to cut up and bone is the large carcase. No longer is the objection relevant that the individual joints are too big, for without the bone they can be cut to any size. But if it is to be a large carcase, then it also must be lean and free, both of external and internal fat. The implications really are far-reaching. For one thing if we could sell carcases of an average deadweight of 25 kg instead of 18-20 kg, it does not take much mathematical ability to work out that the total output from a given national flock of ewes is significantly increased.

This would be of particular importance during the winter months. Hoggetts carried over into the New Year and finished on roots could go a long way to replacing the supplies now coming from New Zealand. Indeed as they do at the present to some extent. But it could be much more so. The basic problem we should have to face would concern a change in selection and breeding of the Longwool crossing ram. For it is the wether lamb by-product of the production of the halfbred ewe lamb which fits this job perfectly – except that the Leicester crosses tend to produce a carcase that is over liberally covered with internal fat.

(3) *Breeding of the Final Crossing Sire*

I have already dealt with this subject in the earlier chapter on breeding, but it is so important for the future that it merits further emphasis. The last fifty years has seen precious little progress in the selection of our meat rams. Based on visual appraisal of conformation, often with overprepared show sheep, the tendancy has been to breed for fat and for the ability to finish on expensive hand-feeding. This bears absolutely no relation to everyday reality. Hopefully, we are now on the threshold of being able to use such equipment as the Ultrasonic Backfat measurer in our selection procedures. This, allied to payment systems based on rewarding commercial quality at the abattoir and penalising fat, could have swift effects on the quality of our ram breeding.

(4) *All-the-year-round Production*

If New Zealand lamb were to become scarce or to be on offer at a much higher price, its present virtue of providing lamb on the British market when home-produced lamb is in short supply could be seriously compromised. Could we fill the gap, as indeed the French do, by producing out-of-season lamb born in October and sold from Christmas to Easter? Technically, of course, we could perfectly easily do this by adopting the same methods of using vaginal sponges and PMS injections to bring the ewes into heat out of season. Perfectly easy to do. The question mark must be whether it is sensible economically. The answer of course lies in the relationship between the value of the lamb at sale and its cost of production, which involves substantial amounts of hand-feeding. I must say that I think it is highly doubtful and I would much prefer to see us developing sensible systems of finishing hill-based store lambs during the winter to meet the need. Indeed the system is already well established – it merely needs extending on to a larger side.

(5) *Ewe immunisation – Fecundin*

At several points in the book I have discussed the importance of higher lambing percentages as a factor in increasing profit. There are two basic reasons why the national performance is so low – sub standard management and the use of stock of low genetic potential. Thus the prescription for a cure is simple to state – although perhaps not always so easy to put into practice. Better management – nutrition and health, and the use of more prolific stock e.g. the Cambridge. Better management can be put in hand immediately and you can expect to see the results in a year – and in any case is absolutely fundamental to progress in any other way. It is totally pointless having lots of lambs born if through faulty management a large proportion of them are lost.

Genetic improvement, on the other hand, either by change of breed or by selection within a breed is a slow business.

There is now another alternative coming up over the horizon – the use of Fecundin. This is a product developed by Coopers and by Glaxo in Australia and New Zealand. Briefly it works like this. Under normal conditions, the ewe produces a steroid in the ovary which restricts the number of eggs shed in each ovulation. Fecundin acts as an antibody to this steroid, thus increasing the number of eggs shed. The bulk of the trial work has so far (to the end of 1984) been carried out in Australia and New Zealand. These first results show a potential increase of around 25 per cent in the numbers of lambs born, with a range of 7 per cent to 45 per cent. It has to be said that these increases are on a low level lambing percentage, and we do not yet know whether similar increases will apply to our much higher levels. UK trials are in progress for 1985 and no doubt we shall soon see.

If it works, and if for the sake of a double vaccination in the first year, followed by a single booster done in subsequent years, we can get a 25 per cent increase in lambing percentage, then it would be safe to predict a ready market for it. The reservation must be: will it work under our conditions and give a similar proportionate increase, and will it risk giving us more lambs than our current management can cope with? Time will tell, but it is certainly an interesting concept which could bring the sort of technological breakthrough to sheep production that we have seen over the years, with cereals for instance.

CONCLUSION

Only five technical changes in the pipeline. The man must be short of imagination if, after writing a whole book, that is all he can come up with. Nethertheless, it is a measure of my belief that if you are seeking drama and excitement in your life and work, then sheep husbandry is not for you. I have tried to make a case for an approach to making money out of sheep that is based on updating old-established principles of husbandry and stockmanship, particularly those that were fundamental to the success of that most successful of intensive livestock systems – the folded sheep flock of the last century. There was much wisdom and shrewdness therein that we have tended to ignore in our search for the modern formula. We would do well to be more humble.

INDEX